U0325658

图书在版编目（CIP）数据

慢行，观风景 ：张剑设计作品选 / 张剑著. -- 福州 ：福建美术出版社，2012.6
ISBN 978-7-5393-2727-3

Ⅰ. ①慢… Ⅱ. ①张… Ⅲ. ①工业产品－造型设计－中国－现代－图集 Ⅳ. ①TB472-64

中国版本图书馆CIP数据核字(2012)第116898号

慢行，观风景

—— 张剑设计作品选

张　剑　著

出版发行：海峡出版发行集团
福建美术出版社
经　　销：福建新华发行集团有限责任公司
社　　址：福州市东水路76号16层
邮　　编：350001
服务热线：0591-87620820（发行部）
0591-87533718（总编办）
印　　刷：福州市德安彩色印刷有限公司
版　　次：2012年7月第1版第1次印刷
开　　本：889mm×1194mm　1/20
印　　张：12
印　　数：0001－2000
书　　号：ISBN 978-7-5393-2727-3
定　　价：85.00元

Savor the Flavor of a Slow Life

慢行，观风景

张剑设计作品选 / 张剑 著

海峡出版发行集团 | 福建美术出版社
THE STRAITS PUBLISHING & DISTRIBUTING GROUP | FUJIAN FINE ARTS PUBLISHING HOUSE

张剑 广州美术学院工业设计学院 副教授

1972——1983 陕西 兴平

1983——1991 江苏 胥浦

1991——1995 就读于无锡轻工业学院 造型系

1995——2009 任教于南京艺术学院 设计学院

2009 至今 任教于广州美术学院 工业设计学院

2006 南京市团委“南京市新长征突击手”称号

2007 团中央光华龙腾基金“2007 中国设计业十大杰出青年”提名奖

2009 第四届中国国际设计艺术博览会“2009 年度工业设计创新人物奖”

2011 中国包装联合会设计委员会“中国设计事业先锋人物奖”

2012 共青团中央直属机构光华科技基金会“科教文化促进项目 CCT 计划青年导师”

网站:【张剑设计】http://www.zhangjian.cn

博客:【慢行，观风景。】http://blog.sina.com.cn/zhangjian1995

送给家人的书

自序

我一直坚持认为设计是一种生活，而非职业。此书收录了继 2005 年出版的《情趣的设计世界》作品集之后七年间一百余件作品，有意回避了我近年来的一些商业设计，并非是我对商业设计的排斥和抵触，而是希望所选的这些更为纯粹的作品能串联起我生活的记录，因此选取作品的衡量标准更多注重“有我”。

七年的时光如猫儿子迈着毛茸的小爪悄无声息地蹭着我的腿从桌边走过,便一下子步入了不惑之年。“不惑”是一模糊的界线，因为分不清自己依旧年轻还是渐将老去，年轻的是保留了以往的执着，但也夹杂着如老者般对过往的回忆。执着如儿时的梦，记忆里理想火烛点燃的那一刻，是在儿时的黄昏美术启蒙老师杨玉敏骑车载着我去县城文化馆学素描的路途中，夕阳火红的余辉将它悄悄点燃。高中几年学画奔波于南京与胥浦两地，虽只有两小时车程，但享受坐长途车看窗外的风景，现在依然记得路途的小河和山丘的模样。南京学画的老师是周炳辰先生,他是著名的版画家,但在我的记忆里他更像一位父亲,最为深刻的记忆是每次画完画欣喜地离开时,周老师都会挺着看不到脚尖的肚子，站在门口目送我离去，微笑的目光穿过架在鼻尖上的老花镜。我记忆里永远珍藏着这个微笑，这个微笑是对一个少年的期盼。

“不惑”是对年龄而言很好的心理定义，“自然而然”取代了年轻时的“刻意”。不强求但也非懈怠。当清晰自己需要什么的时候，很多过于入世的琐碎都可以忽略并舍弃。这会变得简单和轻松，这样的状态在我的作品中希望可以显而易见。读过印度心灵导师克里希那穆提的《静谧之心》之后，很赞同其将诗人、作家、画家等归为“释意者”的概念，设计也如此，我们努力要传递的不是作品本身，而是我们不经意间自然流露的精神及状态。

或许巧合，我为自己的第一本作品集《情趣的设计世界》所写的后记也是在七年前的今夜随性而成。巧合的不止这些，同样在福建美术出版社郑婧的帮助下得以出版此书，依旧是好友朱晓健倾力而成的装帧设计，还是那些老友们为我写的序——这七年似乎又定格在了同样的记忆节点。然岁月改变了每一个人的轨迹。我们都甘愿承受着被生活改变，寂静而非麻木。正如我希望我的作品在简单中能传递浅淡的思考和感动，这样的浅淡或许就是修行。

一路走来，不仓促，是带着记忆在此呈现；并也将一路走下去—— 慢行，观风景。

2012 年 4 月 29 日 广州

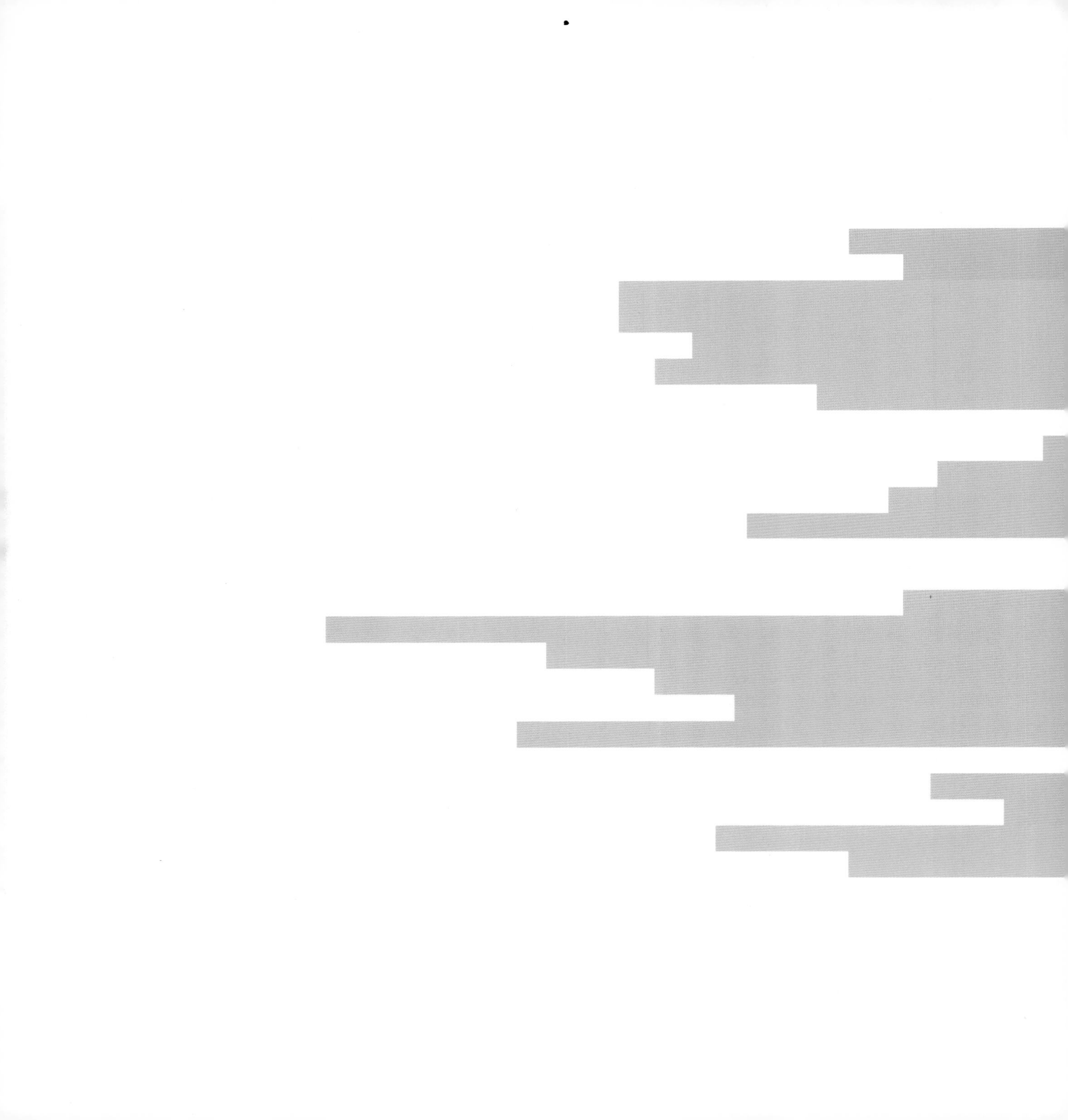

听闻老伙计出这本新的个人设计专辑，祝贺！

一晃经年，我们都已年近不惑，偶尔还会学着少年人那样来点儿感慨。拥有更多的同时身心有些老去，脸上刻上中年人的城府，那是年龄、成就、压力、感悟、经验、疲惫杂糅交织成的复杂的表情——尽管我还在脑海里反复出现张剑老兄当年的风发意气。

写这样的东西，我回忆起很多与张剑作为同学、同事、挚友的许多往事，和张剑在一起是松弛而愉快的，那是至亲伙伴加学术同行的亲密感觉，我们在一起学、一起玩、一起工作，“指点江山，快意江湖”，往事还是那么清晰，越是清晰，越会引发我对现实的唏嘘和感慨。因为同样在这几年经历了几件人生重大的变故，我能想象张剑在设计中的兴奋、执着、坚持、敏锐，也感受到他设计之外的种种复杂的情绪。从上一本个人专辑到这本之间，发生了很多事，首先是这位老朋友专业的精进成熟，那种恣意灵变的创意想象，伴随着他性格中那种随和、玩味、游戏的情趣一直延续着，浸透了他的生活，而生活中的好些事又跌宕突兀，几乎像肥皂剧脚本一样地改变了张剑的许多命运。这些变化，拉出了大大的长长的弧线，长得就像南京到广州的距离——幸好，张剑仍然在这些激变中保持着一个设计者的单纯，单纯的目标、单纯的手艺、单纯的心智、单纯的情趣，单纯的张剑在这个不单纯的年代努力、争取、坚持，有所得、有所失、有所思所感，所以有了这本新的个人设计专辑。我一点都不惊讶这本新书的诞生，它应该是水到渠成的，就像寒冬过后一定是暖春。我有些迫不及待要和他一同享受这春天的温暖了！

我无意对这位老伙计的设计作品做专业的点评，也对他日益增多的成就和获奖并不是特别关注，但这一切却让我感动！偶尔浏览他的网页，就会看到许多新的作品诞生，张剑的创意设计有这样一个好处，即使只是看一看，想一想，也能让我如饮醇酒般地获得心灵的满足，那种快乐细微、恒久、奇妙、温暖——我眯缝起眼睛，仿佛看到他在无数个夜晚伏案工作，电脑屏幕和他的眼镜闪着同步的光斑，几只或茁壮或老态的猫咪忠实地依偎在他的身边，张剑带着他骨子里的老派知识分子般的理想主义，满怀坚定的信念浸淫在南国工作室暖色的灯光下。这情景定格了，让我们这些同是创作人的内心惺惺相惜引发共鸣。这个平时在生活里爱开玩笑，生活有时也和他开几个玩笑的老伙计，能够一如既往地保持下来，真的只有坚持、自信、隐忍才能做得到；真的只有怀揣真诚和善良才能做得到；真的只有像他所说的“简简单单做人，干干净净做设计”才能做得到。设计给张剑带来成就，设计给张剑带来快乐，张剑的设计也给更多人带来快乐与不尽的感悟。

我喜欢看到他真挚开怀的笑容，祝我的老伙计能永远快乐。

挚友 吕江

2012 年春于南京

不惑了吗？

你知道人生最痛苦是什么吗？是在读书、工作、生活中碰到和你同名同姓的人，总能闹出不便和笑话。你知道人生最最痛苦的是什么吗？是有个同名同姓的人每次出书都要让你给他写几句，根本不考虑我现在连给自己写几句的时间都没有。

我和张剑同岁同名（音），于 1996 年南京艺术学院现已不复存在的东楼走廊内结识，我承认我当年私下里接受了他的馈赠——一只耍酷的“zippo”打火机。为了这一贪念我已付出了十六年的代价。1999 年我离开南京去了上海，张剑依然在南艺教书。2008 年我去了北京，张剑也于 2009 年离开南京去了广州。十几年里，我当过他结婚时的伴郎，接待过他搞调研的学生，在电话里聆听过他的倾诉与无奈。虽然相隔渐远，却联系不断。我想这就是朋友。

人生就是一次时空的迁徙，我们都是时空的过客与赌徒。赌徒一定有输赢，作为过客一切终将被时间淹没。回望来路，仍然清晰的青春岁月，只是你我已过不惑之年。我与张剑近年见面的机会渐少，但电话常有，他年轻时说话给人松垮、尖刻、嘻哈的印象，似乎时间在改变一些表象。表象源于他对人生的态度是乐观的，甚或是玩世的，而改变是因为生活毕竟是艰难和苛刻的。这与时代无关，与成长有关。朋友，你不惑了吗？我更加疑惑了。

张剑又来电约我写点东西，我想我只能散记一点生活，对于艺术越研究就越敬畏，而对于评论却越看越觉得迷惘。我想强调的是任何艺术形式都应更多体现人文价值，这是艺术生命力的本源和终极目标。绘画与设计也应当是这种大文化的高华浓缩与精微折射。朋友，我想你应当和我有着相同的价值判断。

书，是人生最好的记录，是来路的真实印记，也暗指了去路的方向。

祝福你，我的朋友。

张见

2012 年 3 月 1 日于上海

目录

趣味家具 Interesting furniture

家居小产品 Household products

花的故事 Story of flowers

工作的周边 Work ambience

生活与概念 Life and concepts

照明与灯具 Lighting and Lamps

创新工具 Creative Tools

卫浴里的乐趣 Joy of bathing

趣味家具
Interesting furniture

Storage Chairs for Kids
儿童储物椅

reddot design award

【时间】: 2008
【Time】: 2008

【奖项】:
2010 德国红点概念设计奖
2008 韩国仁川国际设计比赛入选奖
2008 江苏优秀工业设计二等奖
入选 2010 中国包装技术协会设计委员会《中国设计年鉴》第七卷
【Awards】:
2010 Red Dot Design Concept Award
Selected Work of Incheon International Design Award 2008
Second Prize in Jiangsu Excellent Industrial Design 2008
Selected by Volume 7 of China Design Yearbook 2010, Design Committee of China Packaging Federation

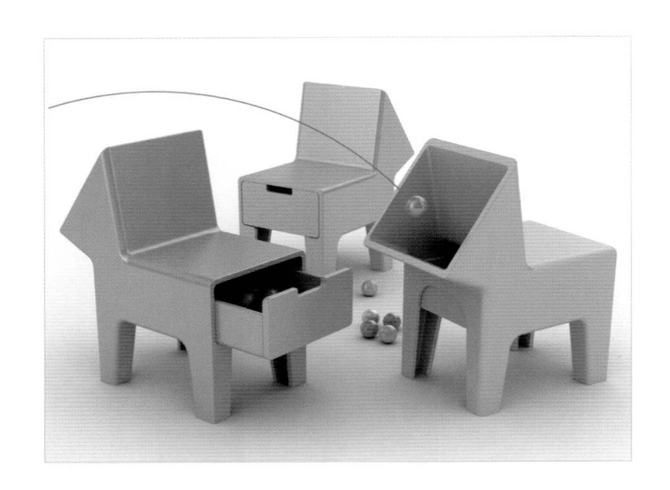

一个张着大嘴卡通造型的椅子，它的功能是用于收纳儿童零乱的玩具。椅子正面座位下方设置了一个很大的抽屉，与椅子背面的大嘴相贯通，儿童可以直接将零乱的玩具与物品投掷在张开的大嘴内，就会落在抽屉里，使收纳的过程变得十分有趣。椅子的材料可以采用塑料，保证了材料的轻便性。

A chair of a comic design is like a big open mouth and its function is to store the scattered toys of the kids. Below the front of the seat there is a spacious drawer, and it' s connected with the big mouth on the back of the chair. Kids can directly throw the scattered toys and items into the big open mouth, and they will fall into the drawer, so that the collecting process becomes full of fun. The chairs can be made of plastic to ensure the portability of the material.

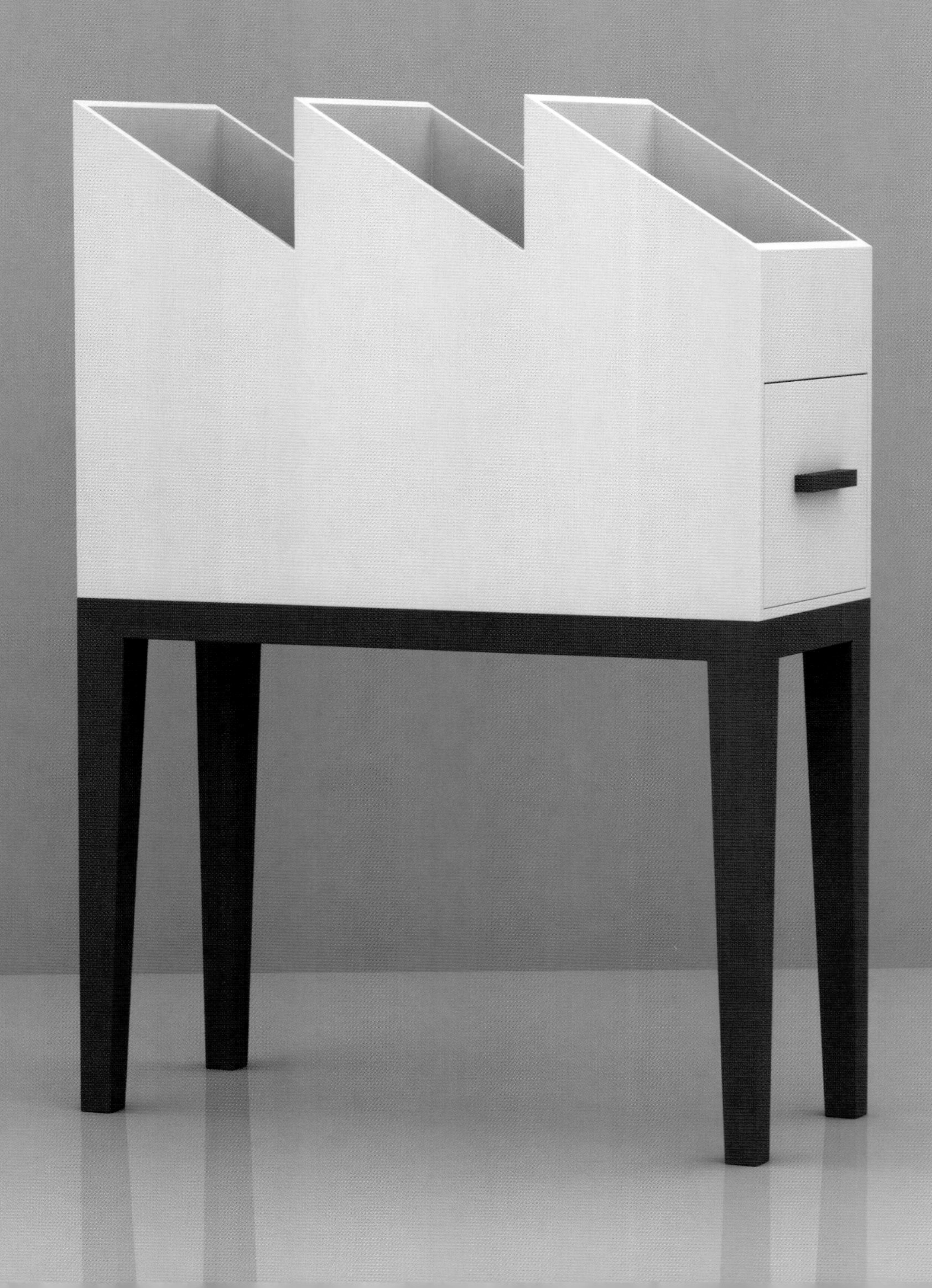

Chimney Series of Cabinet for Sundries
烟囱系列杂物柜

【时间】: 2012
【Time】: 2012

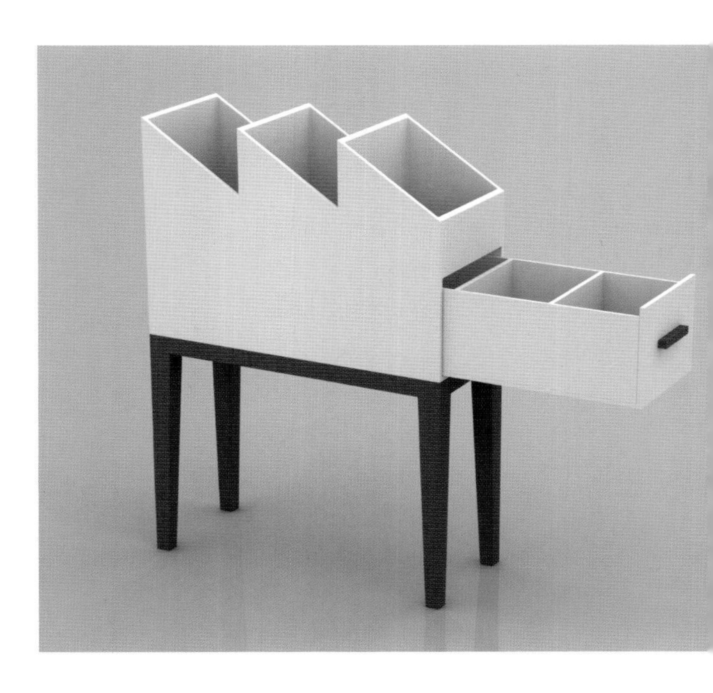

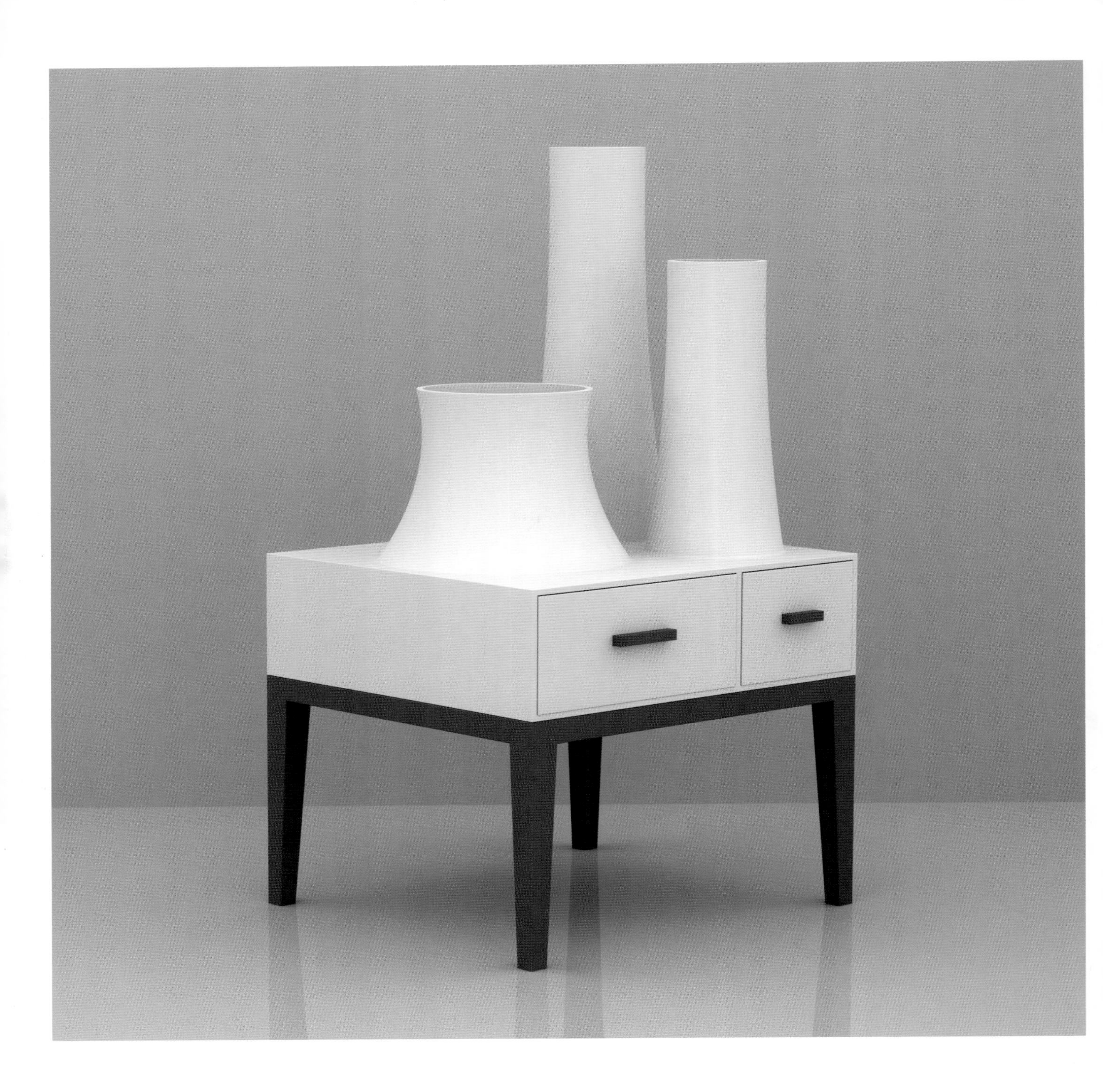

以各式烟囱为造型元素延展的杂物柜，不仅仅为了强调形态的空间视觉落差，产生新颖的家具造型，更主要的是希望实现趣味化的杂物放置方式，每一个烟囱与杂物柜的抽屉相互贯通，而且抽屉按照烟囱的位置设置相应大小的格档，我们可以将杂物直接投掷在不同的烟囱里，杂物则会落在相应的抽屉格档中。杂物柜支架为深色木质材料，烟囱柜体与抽屉为亚光白色塑料。

Cabinet for sundries extended with various styles of chimneys as modeling elements not only aims to highlight the spatial visual difference of forms and produce novel furniture shapes, but also hopes to provide an interesting way for storing sundries. Each chimney is interconnected with the drawers of cabinet for sundries. Moreover, compartments of corresponding size have been set for each drawer according to the location of chimney. We can directly cast sundries in different chimneys, and sundries will fall into the corresponding drawer compartments. The support of sundry cabinet is made of dark wooden material, and the chimney cabinet and drawers are made of honed white plastic.

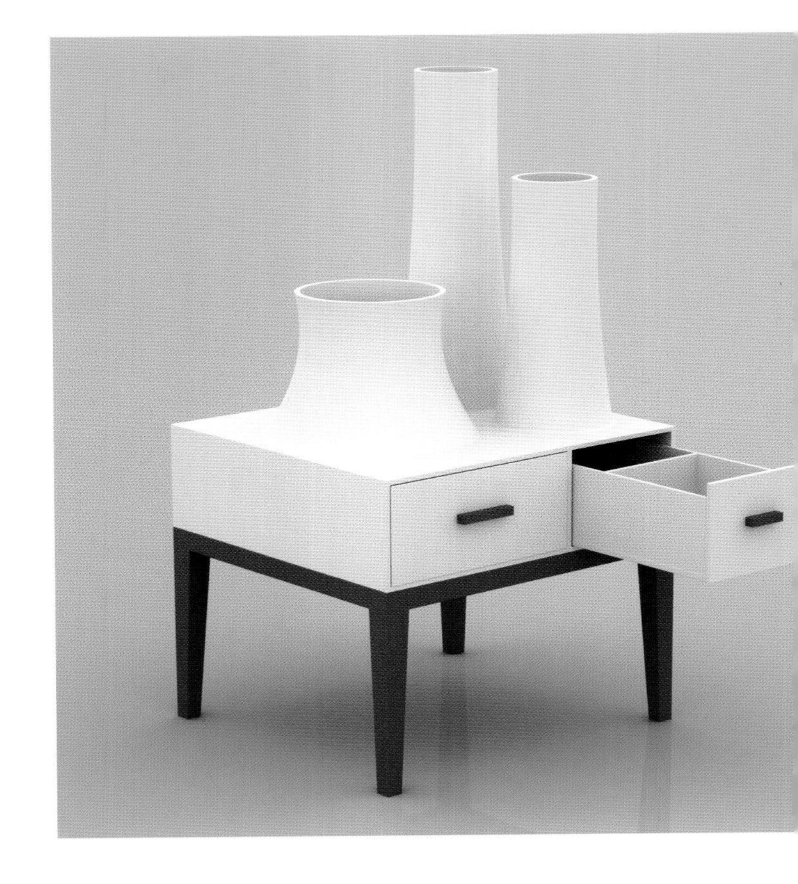

"Here My Home!"

地图桌子

【时间】: 2009
【Time】: 2009

【奖项】: 入选 2010 中国包装技术协会设计委员会《中国设计年鉴》第七卷
【Awards】: Selected by Volume 7 of China Design Yearbook 2010, Design Committee of China Packaging Federation

在茫茫都市里的每个人都能在自己的餐桌的地图上找到属于自己的家，在家的位置插上鲜花是对生活很美好的意愿表达。
餐桌为白色，家庭使用，造型简洁。可以按照不同季节选择花卉进行装饰。
普通的木质餐桌，将所在地城市的地图丝网印在餐桌的表面。
当你在家具店购买这个餐桌时，店员会在你桌面地图所指你家庭的住址钻孔，装上特制的矿泉水瓶接口（矿泉水瓶口统一标准 28 毫米），这样将废弃的矿泉水瓶旋转在桌面底下成为一个花瓶。
普通的餐桌在它的表面印上地图，加上特制的矿泉水瓶卡口，这看似简单的附加设计可以为普通的餐桌带来温馨的感觉。在任何一个城市地图上都能找到属于自己家的位置，并插上鲜花作为装饰。

Everyone can find his/her own place on his/her own dining table in a big city. To place flowers at the location of one’ s home is an expression of the aspiration for a beautiful life.
With a white color and simplistic design, the table is suitable for home usage. Different flowers can be used for decoration in different seasons.
It is an ordinary wooden dining table with a map of your city screen-printed on top of it.
When you purchase this table in the furniture store, the assistant will drill a hole at the location of your home on the map and install a specialized fitting for mineral water bottles (whose diameter is a standard 28 mm). Then, a recycled mineral water bottle is fitted under the table to be used as a vase.
An ordinary dining table with a map printed on top of it, and a specialized mineral water bottle fitting, this seemingly simple design brings a warm feeling to the ordinary dining table. One can find his/her own home on the map of any city and place a boutique of flowers as decorations.

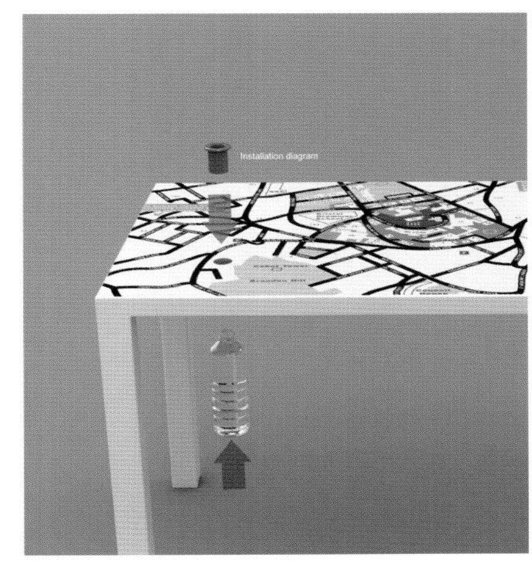

A Kitty Holding On To the Table

趴在桌子边的猫

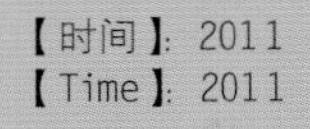

【时间】: 2011

【Time】: 2011

为儿童设计的一款趣味桌子，一只小猫趴在桌子边缘窥视桌子上的食物或玩具。桌子为塑料材质，“小猫”为铝合金材料，并起到支撑桌子的作用，猫的手掌下面有螺纹管，贯穿桌面，与桌面下的螺栓固定。

This is a funny table specially designed for children. A kitty is holding on to the edge of the table, peering at the food or toys on the table. The table is made of plastic material and the "kitty", made of aluminum alloy material, serves as a support of the table. Under the palms of the kitty, there are rifled pipes which penetrate through the table surface and are fixed with the bolts under the surface.

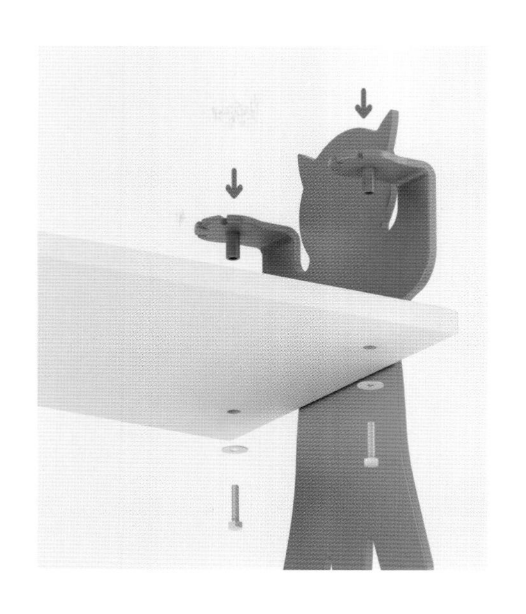

Behind Me

我的背后

【时间】：2005
【Time】：2005

【奖项】：
2007 北京 798 “旋转的硬币” 国际工业设计展优秀作品奖
入选 2008 中国包装技术协会设计委员会《中国设计年鉴》第六卷

【Awards】：
Award of Excellent Work of SPIN COIN 798 International Industrial Design Expositions Beijing 2007
Selected by Volume 6 of China Design Yearbook 2008, Design Committee of China Packaging Federation

由木夹板热弯成型的透空椅子，椅子的前腿由方形扁铝管横穿过椅身并与之固定。椅子的靠背下方向外延伸出一个空间，外部可以作为一书架使用，内部围成的空间可以放包或杂志。但我想，我的猫儿“妮妮”是最喜欢呆在那儿的。

An open-work chair is shaped by hot bending of plywood. The front legs are made of one square flat aluminum pipe, across the chair and fixed with the body. The bottom of the backrest has an extended outward space, which can be used as a shelf. The interior space can be used to put bags or magazines. I guess my cat Nini would love to stay there.

南京的记忆少了物是人非的生活片段，而更多停留于风景与古迹。明孝陵墓道旁的石马经历百年沧桑，记得儿时曾经骑在石马背上照相留影，那么又有多少如我记忆般的场景残存于其他人的记忆的相册里呢？设计这样石马造型的公共坐具，放置于都市公共空间里，骑上它是否能唤起当初的记忆 ？至少能成为记忆相册的见证者。材质为玻璃钢。

The memory about Nanjing is more on sceneries and ancient relics. The stone horses along the path leading to Xiaoling Tomb of Ming Dynasty have witnessed changes of centuries. Do you have the memory as mine of riding on a stone horse and taking a picture when you were a kid? I don’ t know if a public seat designed in the shape of the stone horse and placed in public space in the city would arouse your tender memory, but it would at least be a witness of the album of memory. The material used is glass fiber reinforced plastic.

Stone Horse of Xiaoling Tomb of Ming Dynasty—Outdoor Seat

明孝陵的石马——户外坐具

【时间】: 2005
【Time】: 2005

Combination Stool for Children

儿童组合椅凳

【时间】: 2008
【Time】: 2008

这是一套为儿童设计的椅子与小凳的组合。将小矮椅套在凳子上，可以成为一个高椅子，同时在设计时考虑到儿童尺度因素，凳子和椅子可以单独使用。儿童也可以坐在矮椅子上，将较高的凳子作为书写的小桌子，作为一个较好的书写组合桌椅。材质选用塑料，便于儿童搬运。

This is a set of stool designed for children. When the low chair is set on the stool, it becomes a tall chair. Meanwhile the children' s sizes are considered in the design. The chair and the stool can be used separately. Children can sit on the low chair and take the higher stool as a little table for writing. This makes the combination a quite good writing stool set. Plastic is used as material, easy for the children to move.

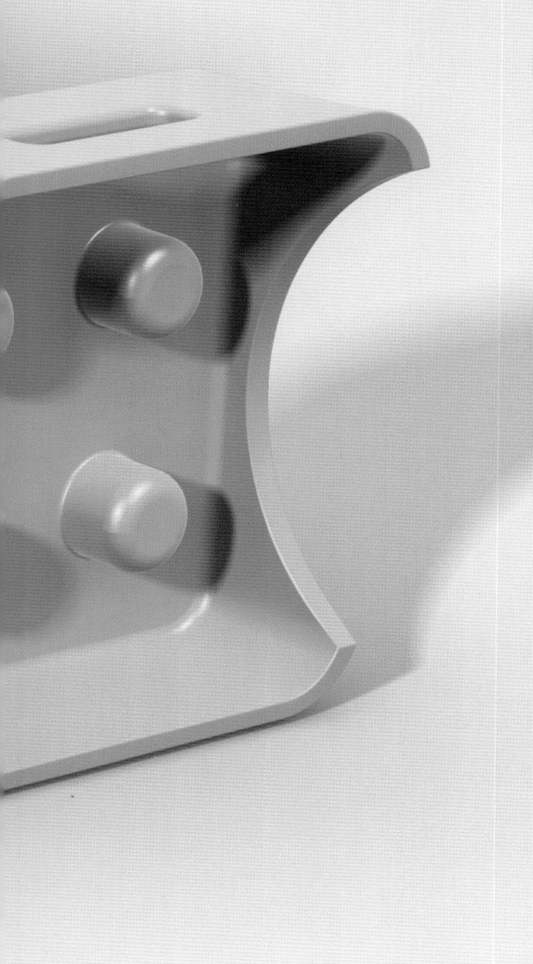

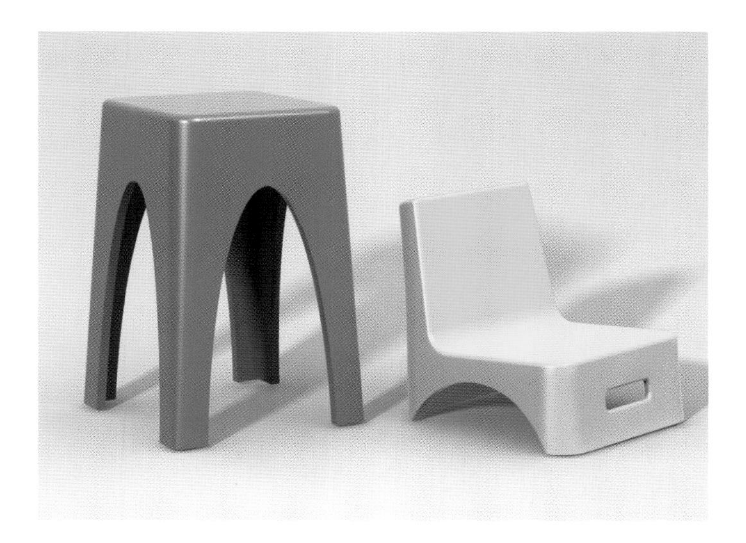

A Stool with Crutch

插拐杖的凳子

【时间】：2011
【Time】：2011

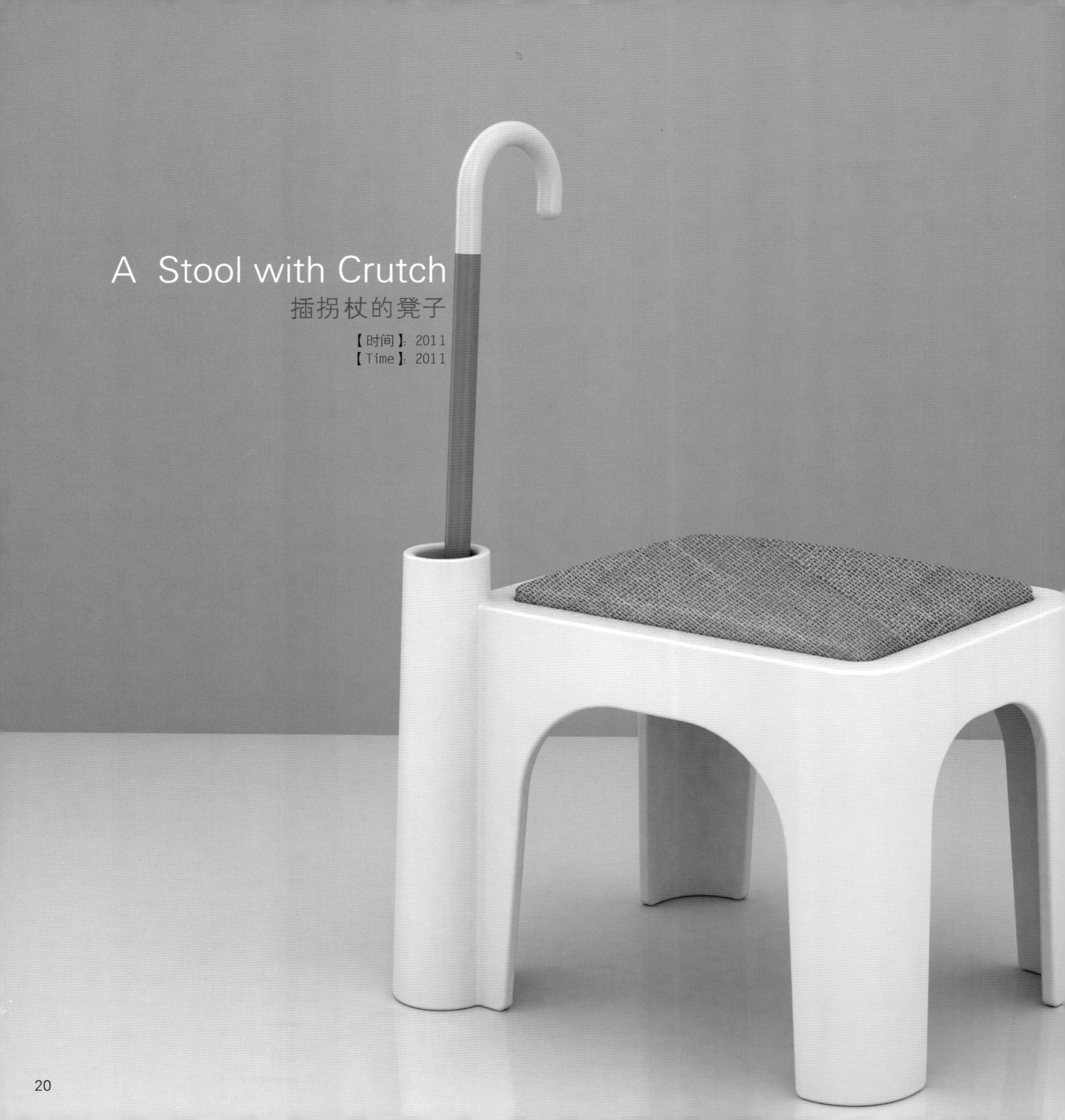

设计这款小凳是对公共坐具的改良，将小凳的右前腿改为圆管造型，这样便于老年人在公共场所休息时放置拐杖，同时拐杖以这样的方式插入凳腿也便于老年人起身时撑扶。当然，下雨天的时候凳腿圆管也可以用于放置雨伞。此设计尝试解决公共场所坐具与使用者行为之间的某些配合关系。

The design of this little stool is an improvement of a public seat. The front right leg of the stool is changed to a tube so that the old people can put the crutch into the tube when they rest. At the same time, the crutch is inserted into the leg of the stool in such a way that the old people can use it as a support when standing up. Certainly, the round tube can also be used to put umbrellas in rainy days. This design attempts to solve some problems of cooperation between the public seats and users.

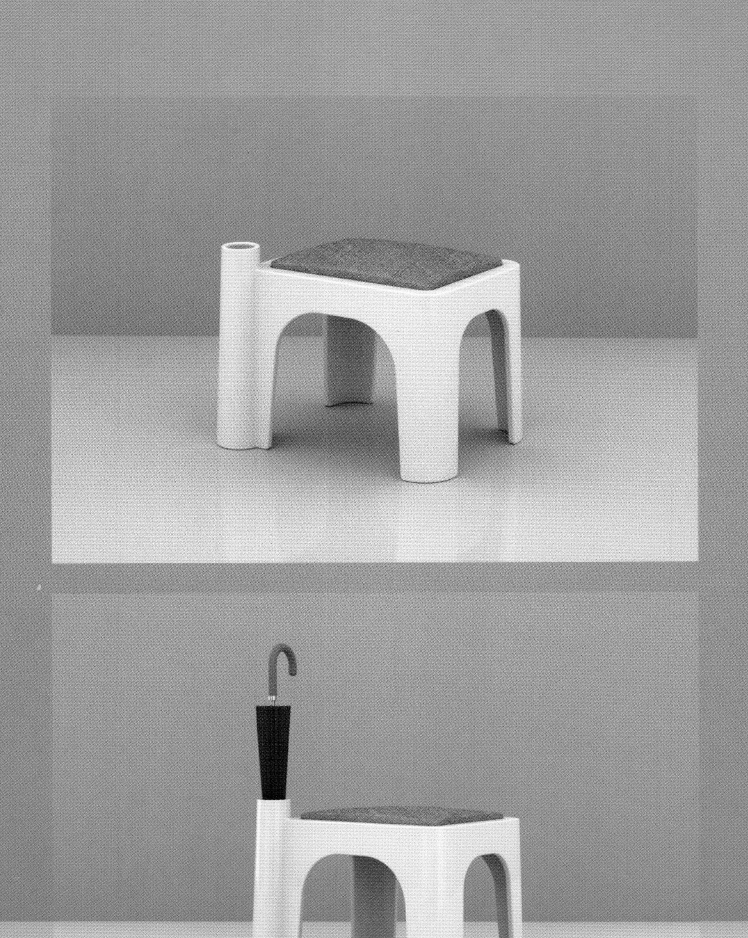

A Table with Scale

带刻度的桌子

【合作】：马莲莲
【Cooperated by】：Ma Lianlian

【时间】：2009
【Time】：2009

【奖项】：入选 2010 中国包装技术协会设计委员会《中国设计年鉴》第七卷
【Awards】：Selected by Volume 7 of China Design Yearbook 2010，Design Committee of China Packaging Federation

这是与研究生合作的为2009年以“尺度”为主题的教学研究作品展的系列作品之一。生产桌子的同时，在桌子表面边缘印上刻度，这样当使用者在工作学习的时候不需要配备额外的尺子，直接在桌子表面进行测量，方便实用。

This is one of the series of works in cooperation with a postgraduate student in the teaching research work exhibition named Scale in 2009. When the table is being produced, the surface edge of the table is printed with a scale. Users will not need to prepare extra rulers while they are working but directly measure on the surface of the table, which is convenient and useful.

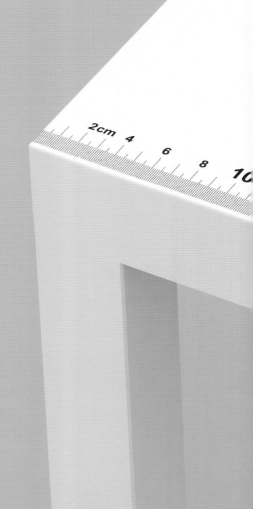

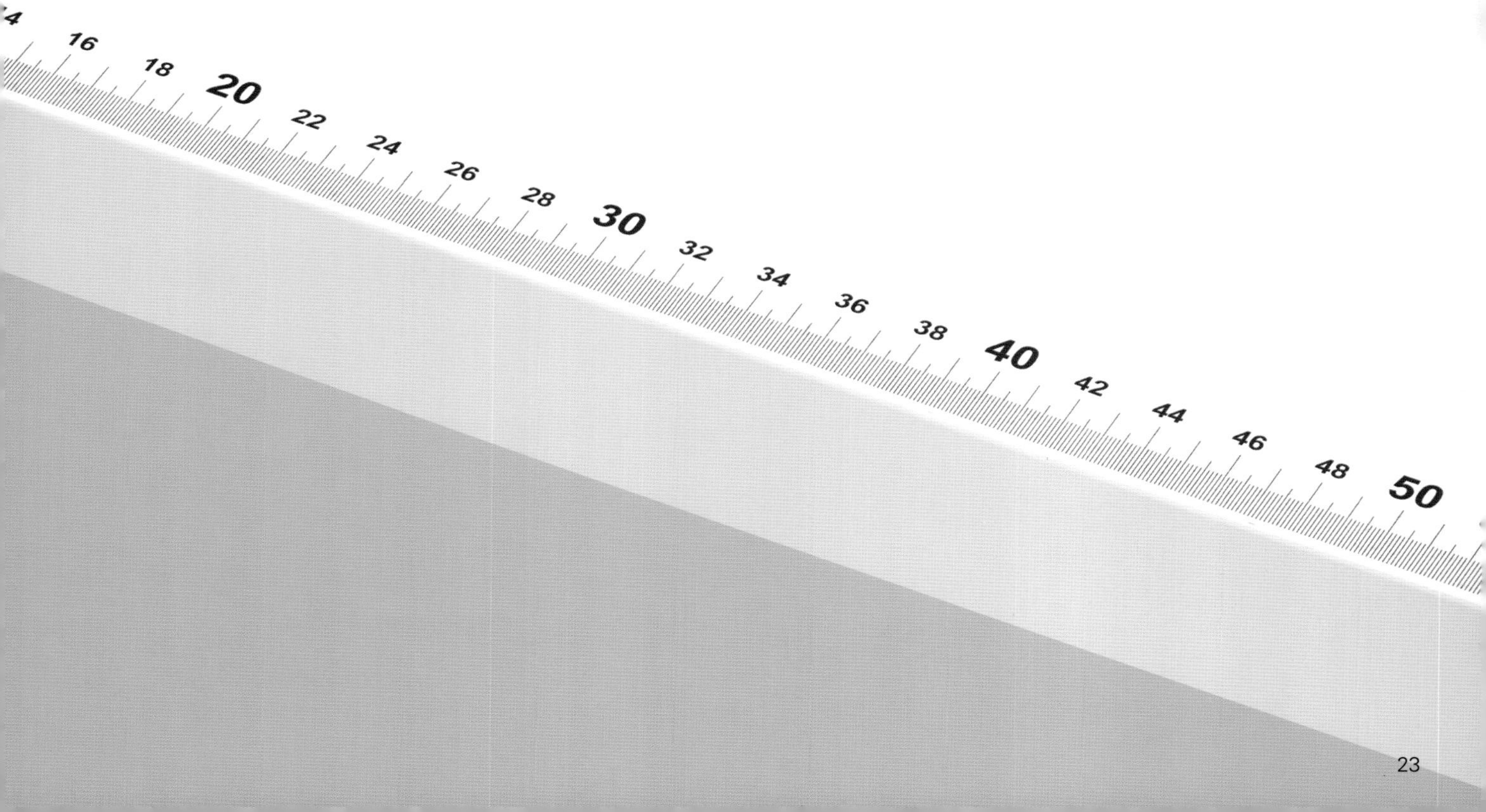
16 18 20 22 24 26 28 30 32 34 36 38 40 42 44 46 48 50

The Side

侧面

【时间】: 2012
【Time】: 2012

为就餐时的行为而做的简单设计。在椅子靠背侧面缝上一枚纽扣，将餐巾布一角开口并固定在椅子的侧面，便于随手使用。

This simple design is for eating behavior. Sew a button on the side of the chair back, and fix a napkin on it. Easy to use.

家居
小产品
Household products

Table–edge Fruit Plate
桌沿边的果盘

【时间】: 2009
【Time】: 2009

一个装水果的椭圆形塑料果篮。将它的上部开一个与普通桌面厚度相当尺寸的水平槽，这样果篮可以卡在桌子的边缘而形成一个环形放置水果的果盘容器，这件作品很巧妙地将果盘与果篮结合在一起。
在结构方面：为了加强果盘卡在桌子边缘的牢固程度，特意将果篮容器与桌子衔接的那个面设计为直面。材料选用塑料，色彩丰富。

An elliptic plastic fruit basket, with a horizontal opening of the same thickness of an ordinary table in its upper part, can be slid and stuck onto the edge of a table to form a circular fruit plate container. This article combines fruit basket and fruit plate in a delicate manner.
In terms of the structure, the surface of the fruit basket which is in contact with the edge of the table is designed to be non-curved, so as to enhance the firmness when it is stuck to the edge. It is made of plastic and rich in colors.

Vessel-shape Fruit Plate

货轮果盘

【时间】: 2009
【Time】: 2009

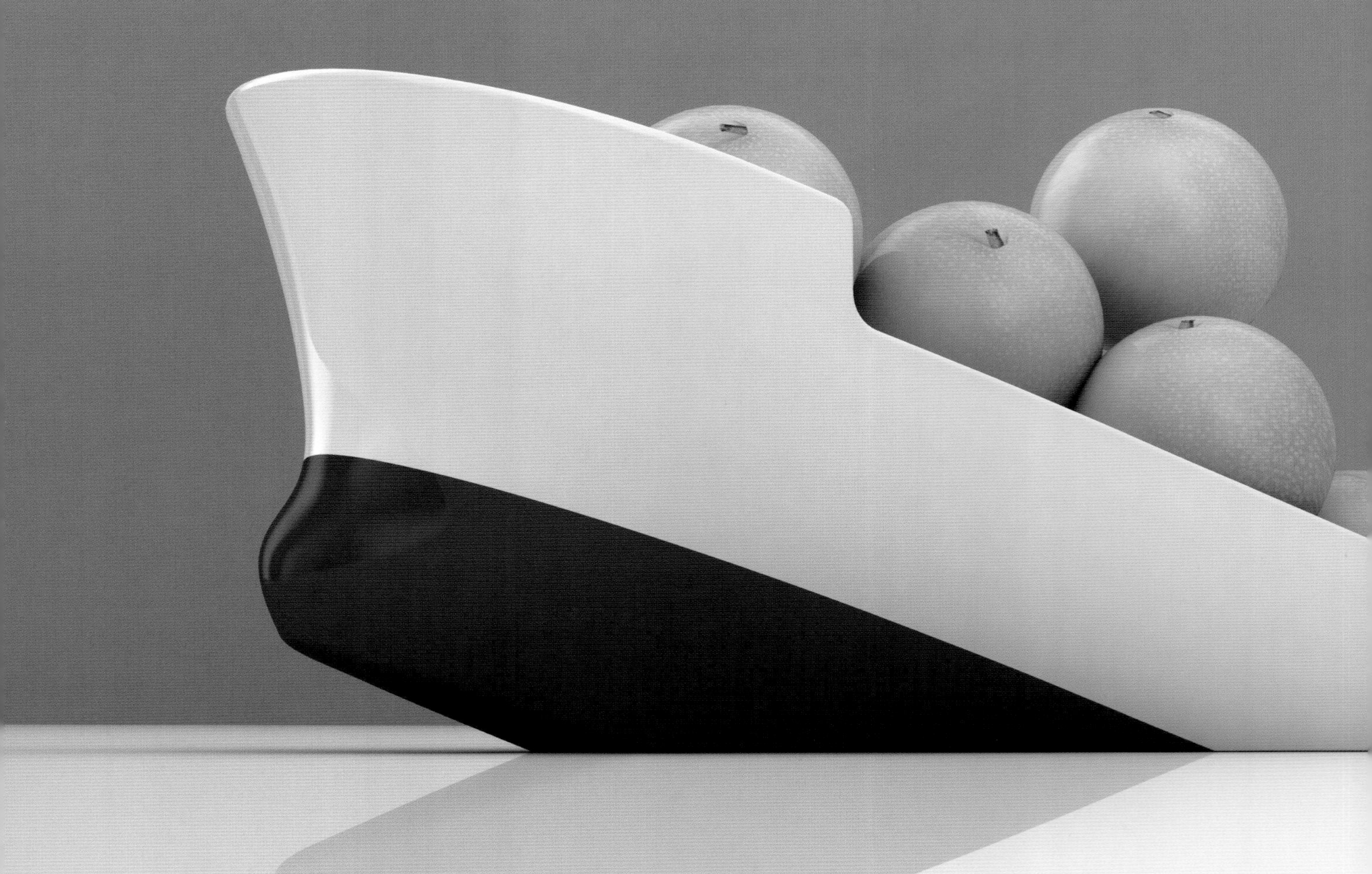

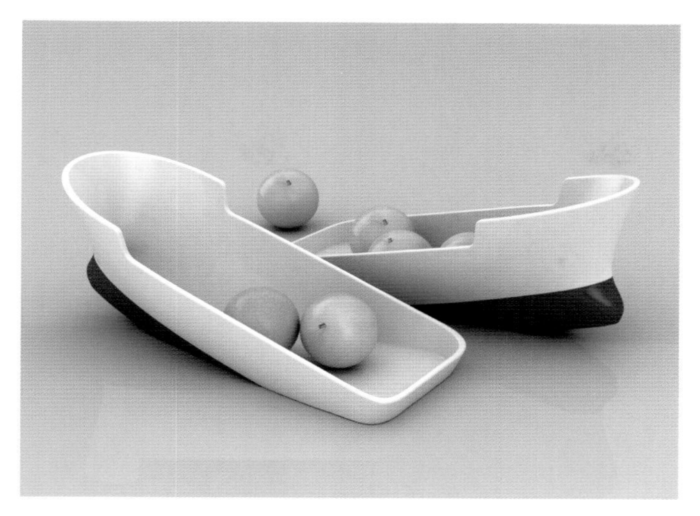

载满水果的货轮，已经承载不了水果的重量几近倾斜颠覆，倾斜的货轮果盘放置在桌面上，桌面好似海面，从而形成一个趣味的场景。

A vessel full of fruits is almost going to turn over due to the weight of the fruits. It makes an interesting scene on the table, which looks like an ocean.

A Fruit Plate Ring

圆环果盘

【时间】：2009
【Time】：2009

器皿可以是有型的也可能是无形的，心理上的有效指示可以达到有型产品同样的功效，一个亚克力圆环放置在桌子表面，其中堆满水果，起到框选的果盘作用，透明彩色亚克力环形表面刻有镂空纹样，不但起到装饰效果，同样提示着环形框选的存在，而这种“存在”的装饰强化，更加强调了框选果盘的功能。

A kitchen ware can be with or without a fixed shape. Effective psychological indication can reach the same effect as a production with a shape. An acrylic ring is put on the table, in which fruits are piled. The hollow cut pattern on the colored acrylic ring not only has the decorative effect but also indicates the existence of the frame in ring shape. The emphasis of “existence” highlights the function of a fruit plate.

Helmet Fruit Holder

头盔果器

【时间】: 2009
【Time】: 2009

如果将水果拟人化，是否可以给它配上一个古罗马战士的头盔。水果戴上头盔就以一个很卡通的战士造型呈现在大家面前，而这样的拟人化设计方式也可以延展到其他类型产品中，得到具有预先设定的趣味情境效果。

If fruit is personated, can it get a Roman helmet? This design makes a cartoon warrior. Such personated design can be used in other types of products to achieve an expected interesting effect.

Horns & Crown—Bottle Cork Design

牛头和皇冠——瓶塞

【时间】: 2008
【Time】: 2008

【奖项】:
2008 意大利" Beyond Silver "国际设计比赛入选奖
入选 2010 中国包装技术协会设计委员会《中国设计年鉴》第七卷
【Award】:
Selected Work of "Beyond Silver" International Design Competition Italy 2008
Selected by Volume 7 of China Design Yearbook 2010, Design Committee of China Packaging Federation

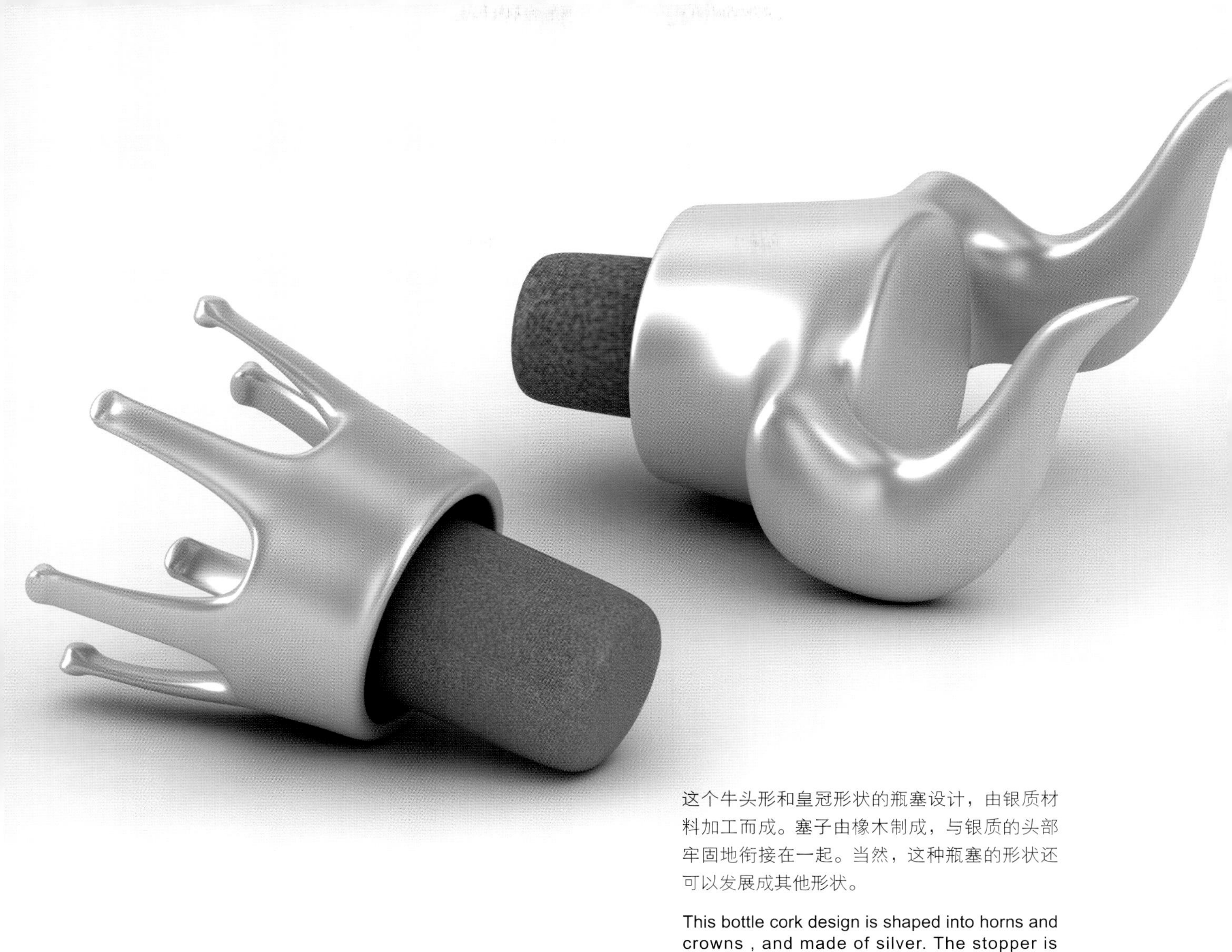

这个牛头形和皇冠形状的瓶塞设计，由银质材料加工而成。塞子由橡木制成，与银质的头部牢固地衔接在一起。当然，这种瓶塞的形状还可以发展成其他形状。

This bottle cork design is shaped into horns and crowns , and made of silver. The stopper is made of oak firmly together with the silver head. Of course, this kind of bottle cork' s shape can be developed into others.

Tumbler Cork 不倒翁水瓶塞

【时间】: 2012
【Time】: 2012
将木塞旋转固定在螺旋钻上
Fix the cork on the spiral drill by rotating it

暖水瓶的软木塞在使用时常会遇到因水瓶里的蒸气长时间蒸热导致拿启时烫手的情况，同时软木塞放置在桌面也会遇到弄脏的问题。设计一款不倒翁原理的软木塞塑料配件，将软木塞通过配件底部的不锈钢螺旋钻旋转固定在配件上成为牢固的整体；塑料配件顶部为空心圆球形，内部配有重金属球，这样当我们使用软木塞时只要握住红色圆球即可完成操作，更具趣味的是，将软木塞放置桌面时，只需将红色圆球随意放置，瓶塞便成为一个不倒翁，不会弄脏瓶塞。

The cork of a thermos flask often scalds hands because of the heat from the long-time steaming in the flask. In addition, when placed on a table, the cork often turns dirty. We have designed a plastic component of cork according to the theory of tumbler. We can fix the cork on the component by rotating it through the stainless spiral drill on the base of the component, so that it can become a firm whole. The top of the plastic component is a hollow sphere, with a heavy metal ball inside. When we use the cork, we can complete the operation as long as we hold the red round ball. More interestingly, when we place the cork on the table, we need only to randomly place the red round ball and the cork becomes a tumbler and stays clean.

Bottle Plug Series

瓶塞系列

【时间】: 2011
【Time】: 2011

以葡萄酒瓶塞作为载体的系列设计，制作一些具有个性风格的瓶塞。在这个系列中通过选取五金店铺中相应的五金配件与瓶塞的使用功能相配合，是尝试以现有材料为出发点的设计方法。同时也试图以各种木质材料为基础，解决木质材料间造型及色彩纹理的搭配问题。

The series of designs have offered some special bottle plugs. Hardware fittings are used in the series to match the function of the bottle plugs. At the same time, wooden materials are also used as the basis to solve the problem of moulding between wooden materials and the matches of colors and textures.

Burning Bottle—Bottle Plug

燃烧瓶——瓶塞

【时间】: 2012
【Time】: 2012

这个瓶塞设计或许有些“疯狂”，如火焰燃烧般的红色方布与木质瓶塞固定，红布与酒瓶的配合好似燃烧瓶，虽静止但却有按捺不住热烈如火的冲动。

This bottle plug design might be a little “crazy” . The flame-like red cloth is fixed with the wooden bottle plug. The combination of the cloth and the botlle is just like a burning botlle,static but with the impulse as hot as fire.

Boat—shape Bottle Plug

船瓶塞

【时间】: 2012
【Time】: 2012

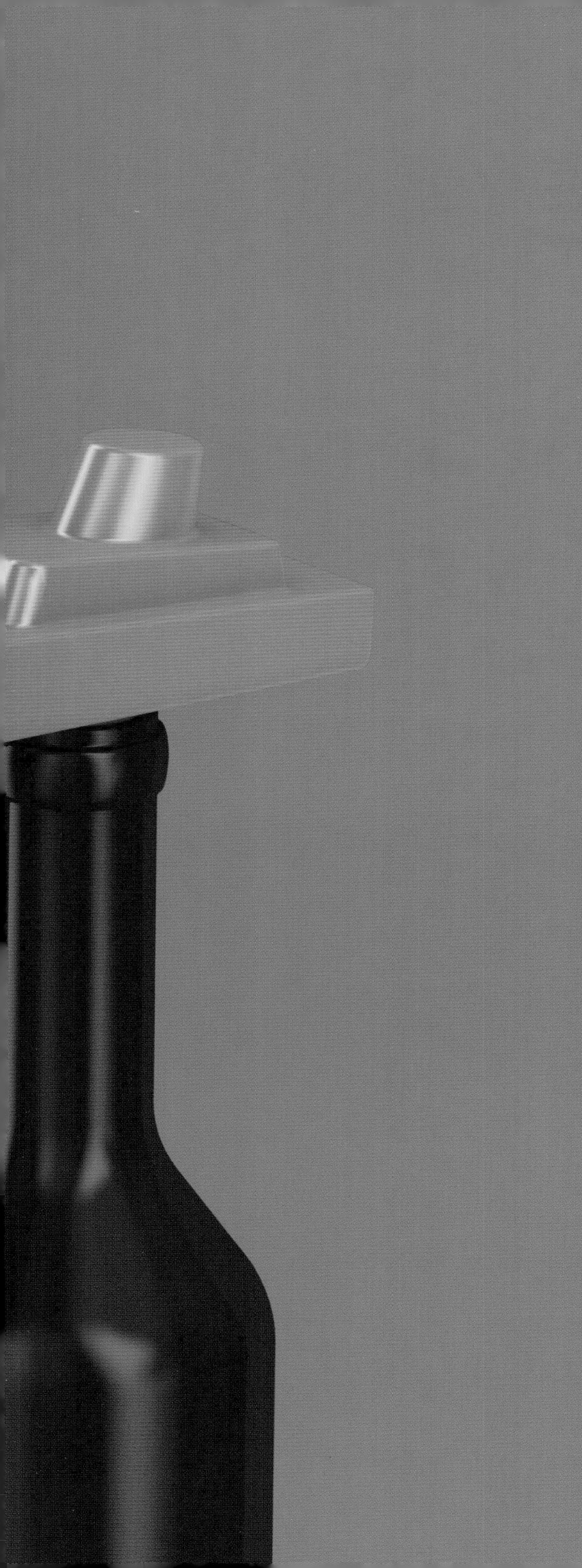

之所以以各类船造型来做瓶塞，不单是装饰，当喝完酒，酒瓶空后，这瓶塞还可以作为漂流瓶。试想，漂流瓶内的祝福与心愿随一艘小船飘向未知的远方是多么浪漫的期盼。当然，将祝福写在纸上装入瓶中之前，要在瓶内装入适量干沙以作配重，这样瓶子才能垂于水中，轮船显露于水面。轮船形态为塑料材质，底部有软木塞与之连接。

The reason for using different boat shapes as bottle plugs is not only for decoration, but also for the purpose of using the empty bottles as drift ones. It would be a great romance to drift the bottle with wishes and hopes. Certainly, before the paper with wishes is put into the bottle, some dry sand should be put into it first to balance the weight so that it can stand in the water and the boat will appear above the water. On the bottom of the boat, there is a cork connected.

Yoghurt Bottle Redesigning

酸奶瓶再设计

【时间】：2012
【Time】：2012

收集家里喝完的酸奶瓶，洗净后为他们设计一系列的盖子，有奶嘴、冰淇淋及牛头造型，选择这些造型的目的是希望能与酸奶瓶原有的使用情境达成贯通的思维联想。这些瓶子可以按照不同的瓶盖造型分别成为橡皮筋、图钉、文具夹等杂物的收纳容器。

After collecting bottles in which yoghurt has been drunk up, you can design a series of lids for them, shaped like nipples, ice creams and cow heads. These shapes are chosen to achieve a cut-through thinking association with the original using situation of yoghurt bottles. These bottles can become containers of such sundries as bungees, thumbtacks and stationery clips respectively according to different shapes of lids.

Drinking—Bottle + Glass
酌——酒瓶 + 酒杯

【时间】: 2005
【Time】: 2005

【奖项】: 2005 "改变包装"国际设计连线全球竞赛最高奖，全场大奖
【Award】: Highest prize in [REDEFINE PACKAGE] International Online Design Contest 2005, The Grand Prize

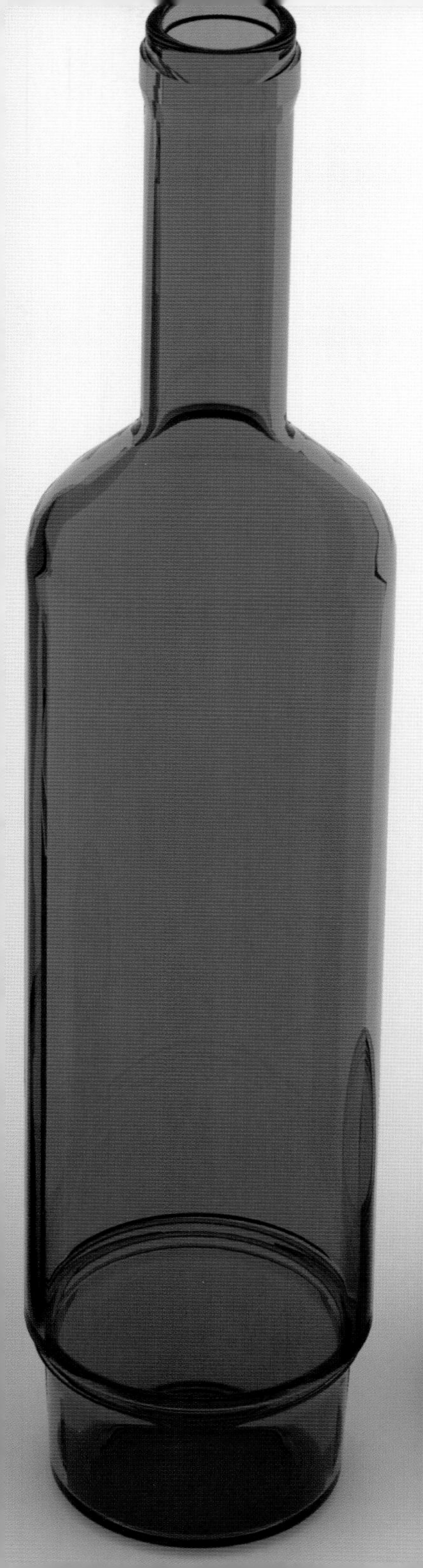

这件作品是参加一个名为“改变包装”的平面设计比赛，并获得全场最高奖项。一个再普通不过的酒瓶，底部配上一个酒杯，表达“自斟自饮”的状态。很多人疑惑这件粗陋的作品居然能获全场大奖，我想，这件作品打动所有评委的应该是剔除了“搔首弄姿”的矫情奢饰之后所显露的本真吧。

This work has won the highest prize in a graphic design competition called “Change Packaging” . It is a very normal bottle with a glass matched to the bottom, expressing the state of “drinking alone” . Many people have the question why such a rough work could win the highest prize. I guess the reason is that it has attracted the referees by showing the innocence and trueness.

正如我在课堂上所讲的，设计活动围绕创意的“原点”展开，任何与之无关的装饰与附加都会带来或多或少的思维歧义，甚至于是矫揉造作。但简洁明了也并非无趣的直白、空洞无物。

As what I have said in class that designs should be carried out around the “original point” of creation. Any unnecessary decoration or add-ons will bring more or less ambiguity, and even artificial sense. However, simplicity does not mean uninteresting or devoid of content.

Drinking—Wine Pots and Cups
酌——酒器

【时间】: 2007
【Time】: 2007

【奖项】: 入选 2008 中国包装技术协会设计委员会《中国设计年鉴》第六卷
【Award】: Selected by Volume 6 of China Design Yearbook 2008, Design Committee of China Packaging Federation

这款酒器是 2005 年“酌”作品的深入设计，也是中国传统酒盅的改造，将酒杯与酒盅结合在一起。酒杯内沿有螺纹，可以与酒盅底部的螺纹旋在一起成为整体。一个酒盅配一个酒杯旨在体现自斟自饮的文人雅士“独酌”状态。

This design was evolved Chinese from the traditional wine-pot' s style.By the combination with the wine-pot and wine-cup, the wine-cup can be revolved and fixed on the bottom of the wine-pot.This design shows the drinker' s relaxed state when they drink with themselves.The wine-pot' s material is porcelain.

Solidification of Ancient and Modern Bottles' Design

古今的凝固

【时间】: 2008
【Time】: 2008

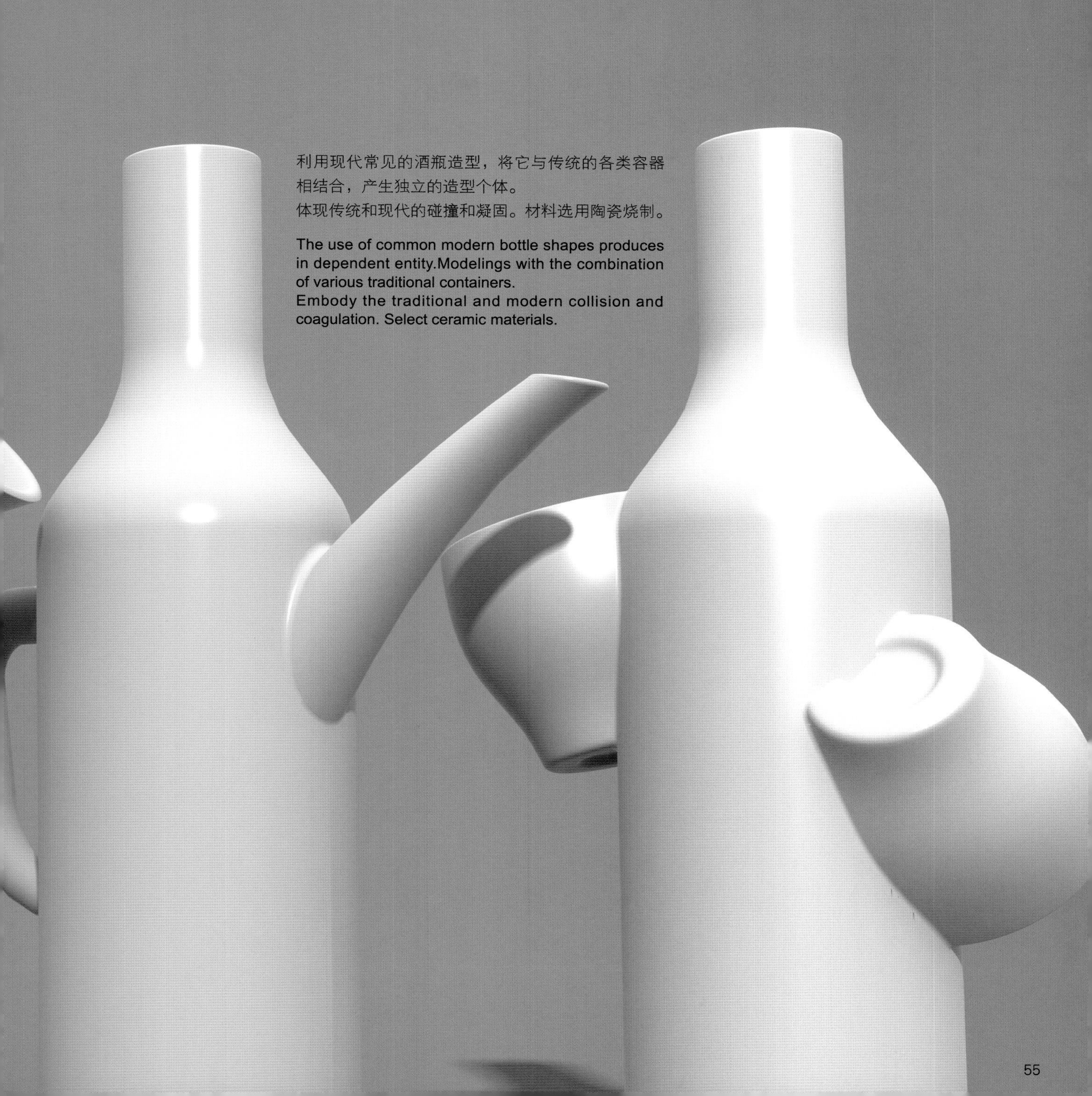

利用现代常见的酒瓶造型，将它与传统的各类容器相结合，产生独立的造型个体。
体现传统和现代的碰撞和凝固。材料选用陶瓷烧制。

The use of common modern bottle shapes produces in dependent entity.Modelings with the combination of various traditional containers.
Embody the traditional and modern collision and coagulation. Select ceramic materials.

Horizontally-placeable Wine Bottles
可以平置的酒瓶

【时间】：2010
【Time】：2010

我们经常遇到这样的问题：葡萄酒或其他的酒类在保存的时候，尤其当瓶子开启后需要将酒瓶水平放置在酒架上，这样才能避免空气进入酒瓶，以此来保证酒的醇香。于是设计了这样的一款酒瓶：将酒瓶的侧面做一个纵向的平面，这样当瓶塞塞在酒瓶上时，酒瓶可以平置于桌面上，保存与放置相当方便。
当然在设计这款酒瓶时候，酒瓶需要比普通的酒瓶稍微高些，或粗些，以保证装酒的容量达到标准。

We only encounter this kind of problem: when storing wines or other alcohol, especially after the bottles are opened. The bottles need to be placed horizontally on the rack in order to prevent air from entering the bottles to ensure the fragrance of the wine. Therefore this bottle is designed to have a flat vertical surface at the side of the bottle, so that it can be placed horizontally on a table with the cork plugged, which is a great convenience to store and place the wines.
When designing this bottle, we need to make it slightly taller or wider than ordinary bottles to ensure the capacity meets the standard.

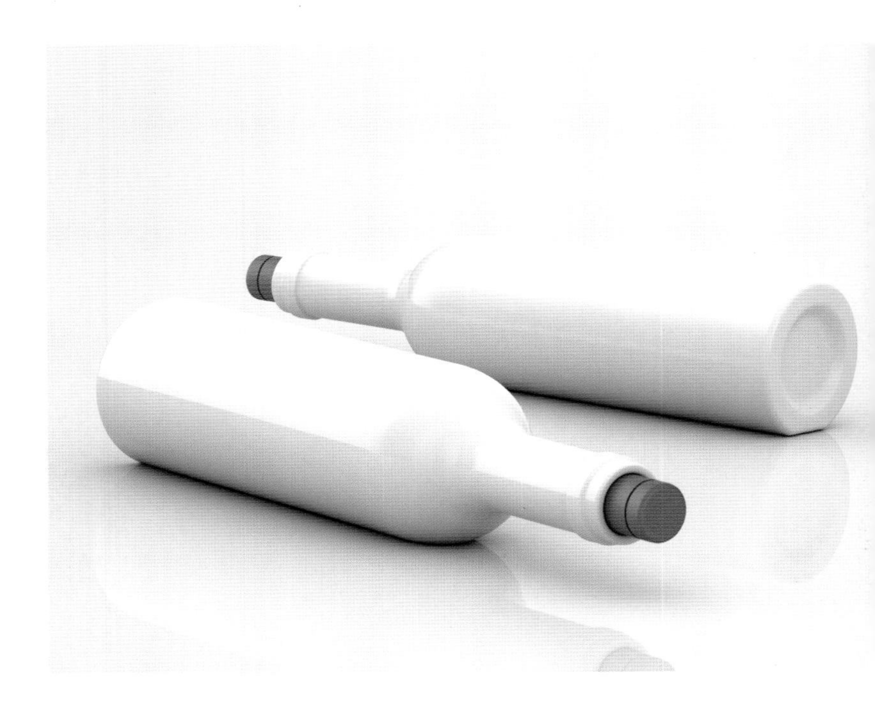

Buoy Scale
浮标刻度

【时间】: 2008
【Time】: 2008

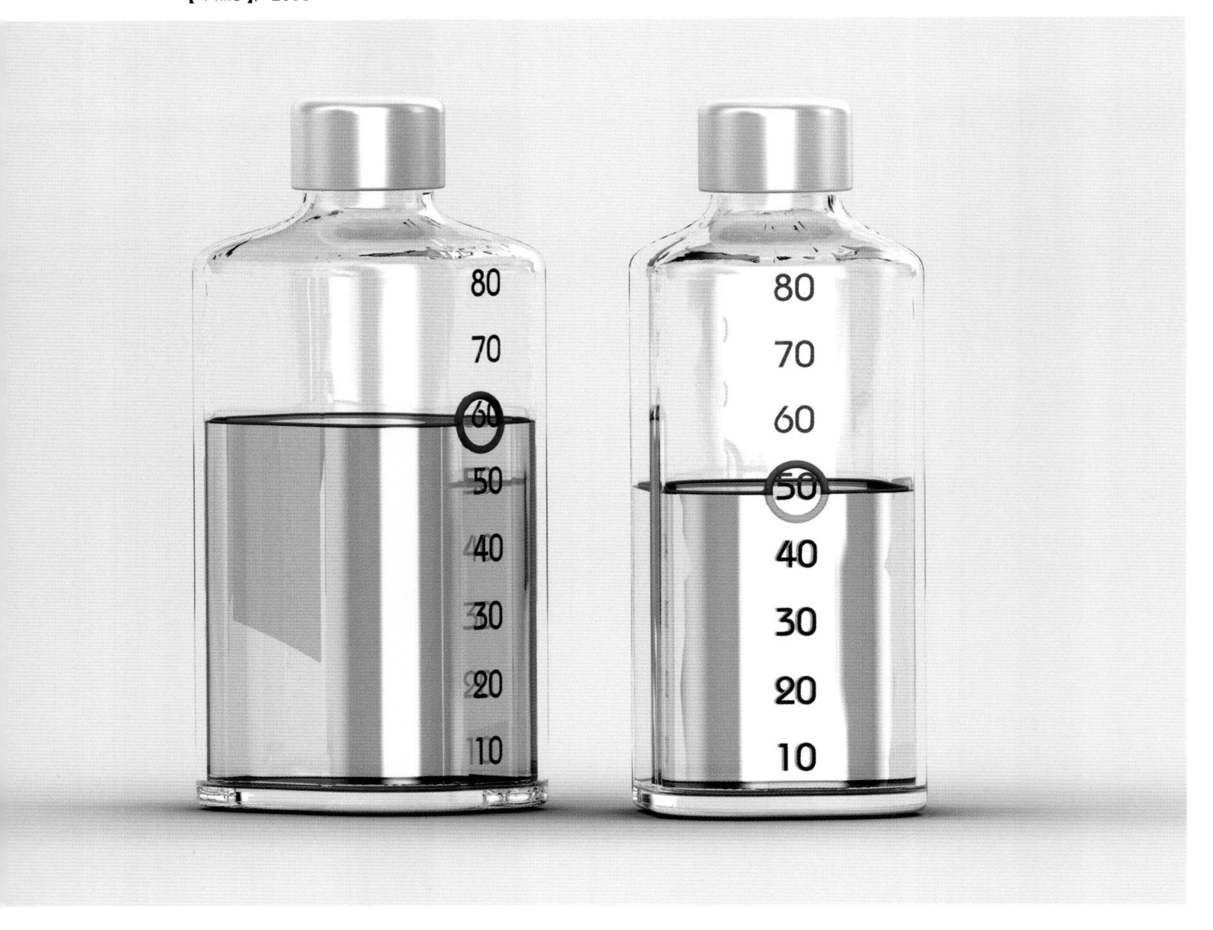

酒瓶内放置一个与酒瓶内径大小一致的无毒空心红色塑料浮标，酒瓶外侧垂直丝网印刷上酒的容积刻度，这样在饮酒的时候，酒的剩余量与飘浮在酒瓶内部的浮标相对应，以便控制饮酒量。

Place a hollow non-toxic red plastic buoy inside the bottle, whose size is consistent with the diameter of the bottle. The volume of wine scale is printed outside the bottle. In this way, as the remainder of wine is corresponding with the floating buoy, it' s easy to control alcohol consumption.

Chimney—Ashtray Series Design

烟囱——烟灰缸系列设计

【时间】: 2009
【Time】: 2009

【奖项】:
2009 创新顺德国际工业设计大赛专业组铜奖
入选 2010 中国包装技术协会设计委员会《中国设计年鉴》第七卷
【Award】:
Bronze Prize of Professional Section of Innovation Shunde International Industrial Design Competition 2009
Selected by Volume 7 of China Design Yearbook 2010, Design Committee of China Packaging Federation

此设计将工业生产的各类烟囱造型结合到烟灰缸设计中，将香烟放置在烟灰缸中，烟雾会从烟囱口飘出，设计师想通过这一系列的烟囱造型传达吸烟对人体的危害就像工业生产中从烟囱排放的对环境产生污染的废气一样的设计思想。材料选用陶瓷，色彩选用白色与黑色以加强对比。

This design molds various types of the shelf chimneys and ashtrays. Put cigarettes in the ashtray, and smoke will drift from the chimney, Through this series ashtray designs, the designers want to convey the hazards of smoking on the human body just like waste gas on the environment from the chimney during industrial production .The material is ceramic. The color is white and black to enhance contrast.

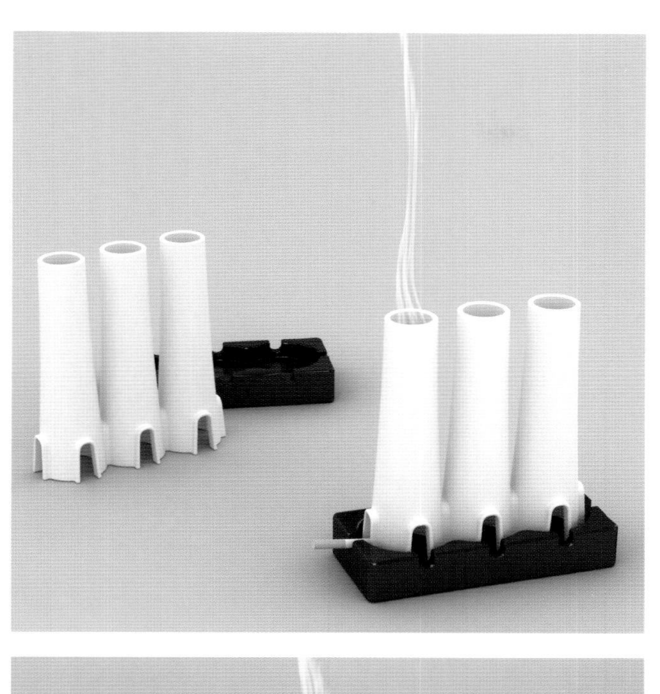

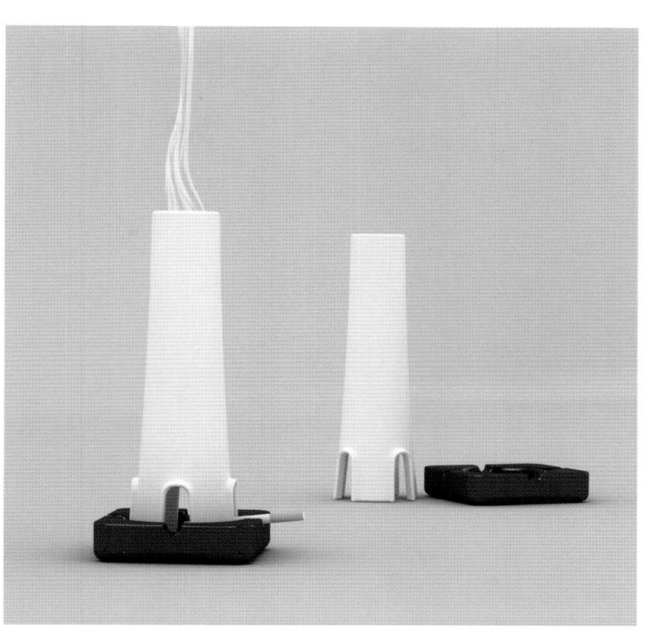

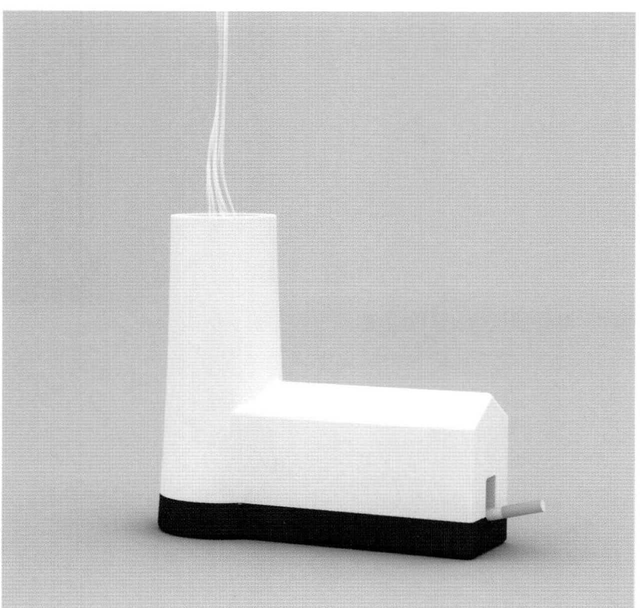

3D Puzzle Home Series

立体拼图家居用品系列

【时间】: 2008
【Time】: 2008

【奖项】: 入选 2010 中国包装技术协会设计委员会《中国设计年鉴》第七卷
【Award】: Selected by Volume 7 of China Design Yearbook 2010, Design Committee of China Packaging Federation

此设计为家居厨房用品系列，贯穿整个系列的造型元素是黑白两色的不同造型的立体拼图，色彩对比强烈生动。

The design for the home kitchen supplies Series. The modeling elements throughout the entire series are in black and white and with 3D puzzles of different shapes, whose color is vivid and with strong contrast.

果盘：拼图本身采用仰角设计，拼上后即可起到卡口的作用

Compote: The puzzle design uses elevation, which makes the compote firmly tight once finishing.

烟灰缸： 拼图造型的各单元，插紧固定后成为烟灰缸。

Ashtray:Fit every parts of the puzzles tightly together to become an ashtray.

调味瓶：四个造型环抱在一起，由于每个造型一样，特意将调味瓶开口小孔的布局做了区分，以便于操作使用。

Cruet：With four identical shapes surrounded together. Purposely distinguish the layout of the opening holes, in order to facilitate the using operation.

蜡烛台：可以拼在一起使用，也可以每个单元单独使用

Candle holder：The candle holders can be put together and used as a whole or separated from each other as a single unit.

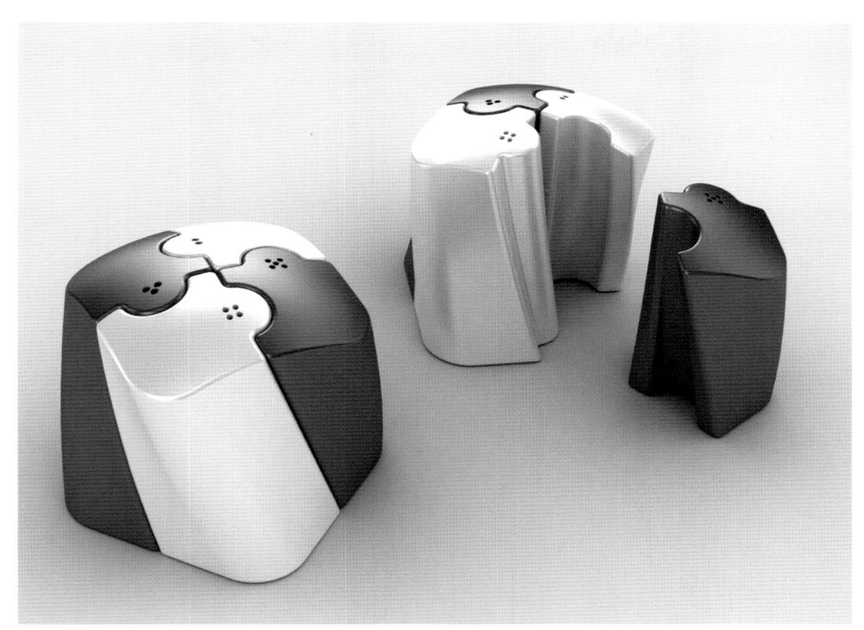

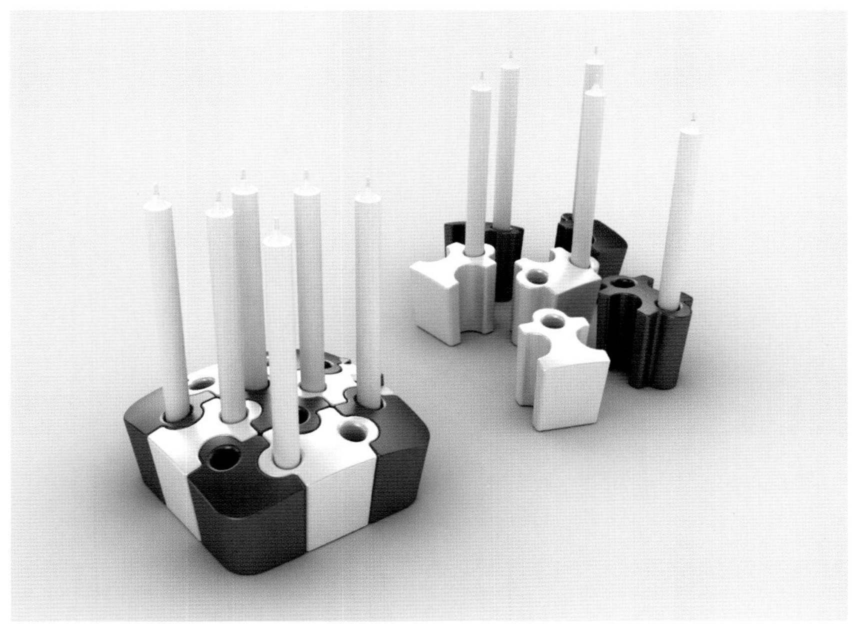

Shredding Container

刨丝容器

【时间】：2007
【Time】：2007

这是一个将刨丝的刨子与容器相结合的设计。在容器的两侧分别设有刨丝的刨槽和用于紧握容器的把手，这样就可以轻易地将各类蔬菜水果刨出的丝收集到容器中。容器材质：不锈钢。

This is a design, which combines a shredder with a container. Two sides of the container have respectively a shredding trough and a handler for clasping the container. Thus various shredded vegetables and fruits can be easily gathered into the container.Material of container: stainless steel

Outdoor Picnic Bag
户外野餐包

【合作】: 于庆庆
【Cooperated by】: Yu Qingqing

【时间】: 2008
【Time】: 2008

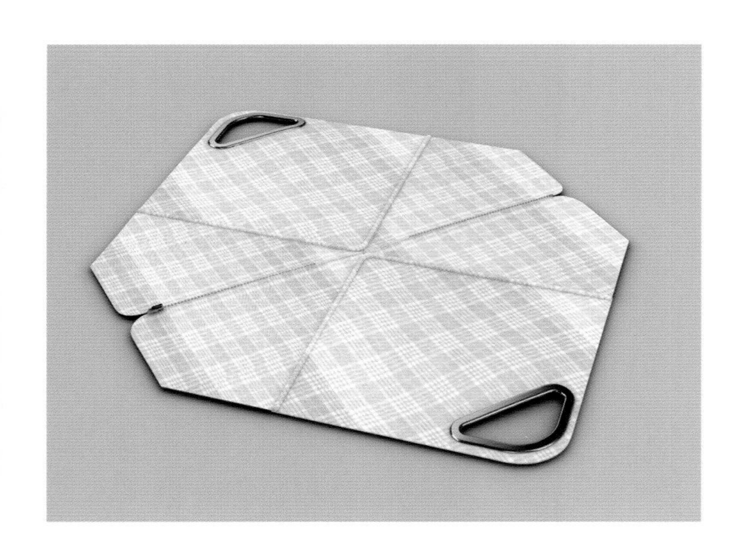

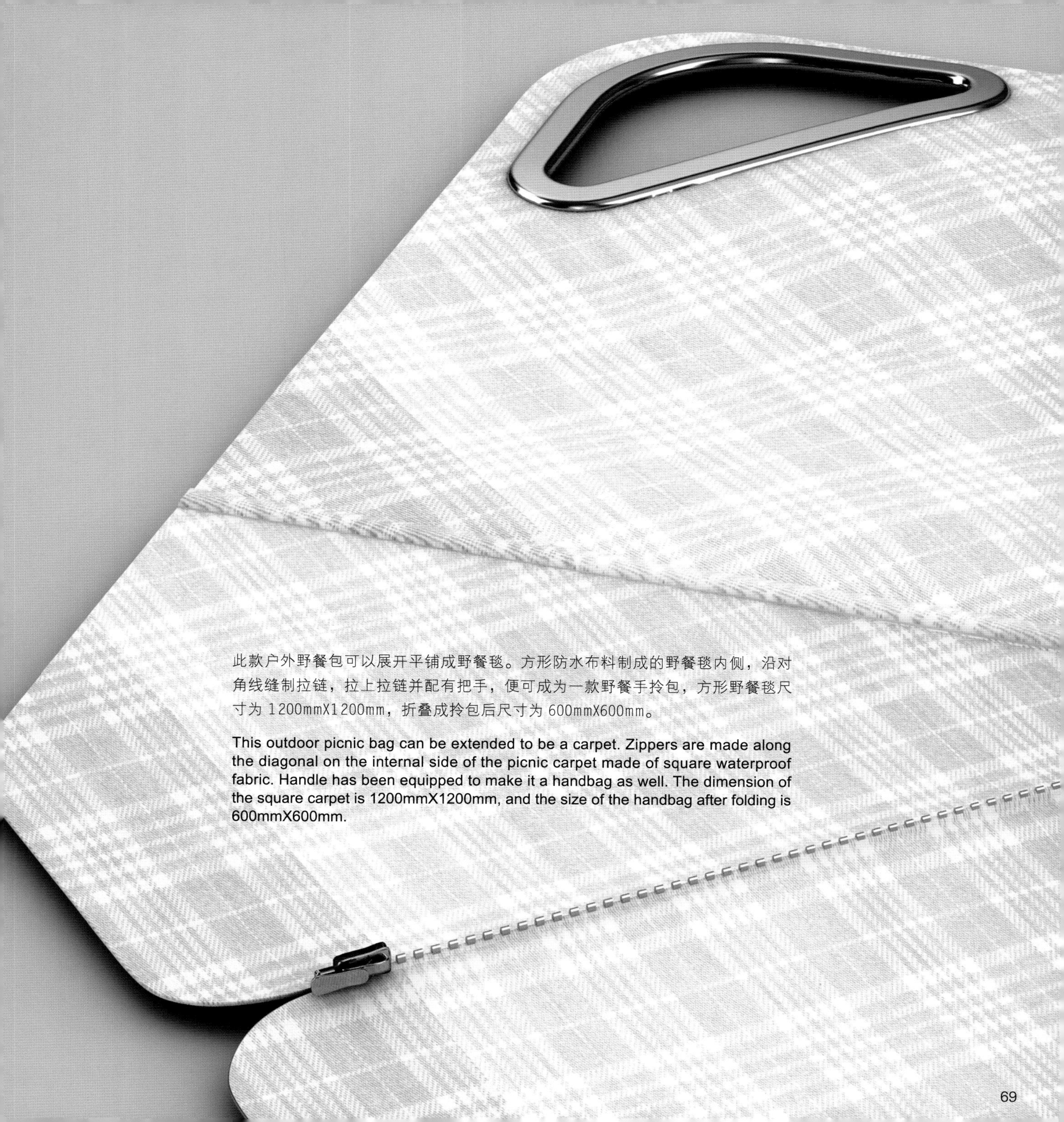

此款户外野餐包可以展开平铺成野餐毯。方形防水布料制成的野餐毯内侧，沿对角线缝制拉链，拉上拉链并配有把手，便可成为一款野餐手拎包，方形野餐毯尺寸为 1200mmX1200mm，折叠成拎包后尺寸为 600mmX600mm。

This outdoor picnic bag can be extended to be a carpet. Zippers are made along the diagonal on the internal side of the picnic carpet made of square waterproof fabric. Handle has been equipped to make it a handbag as well. The dimension of the square carpet is 1200mmX1200mm, and the size of the handbag after folding is 600mmX600mm.

Cup Jointed with Drinking Bottles

与饮料瓶连接的杯子

【时间】: 2007
【Time】: 2007

基于饮料瓶瓶口的尺寸是标准统一的，于是设计了可以直接旋在饮料瓶上的杯子。喝水时每人一个这样的杯子，可以分享同一瓶饮料。
杯子由杯身与杯托两个部分组成，杯身为陶瓷或塑料制成，杯托为木制。当杯子从饮料瓶上取下时，可将其和木质底座旋紧后，做为一个普通的水杯使用。杯托的造型可以做成酒盅的造型，或者其它形状色彩以增加饮用时的乐趣。

As the mouth sizes of drinking bottles are standard and unique, I have designed a kind of cup, which can be screwed directly with drinking bottles. Therefore, if everyone has one cup of this kind, when drinking, they can share drinks in the same bottle.
The cup consists of cup body and cup tray. The body is made of china or plastic, while the tray is wooden. When the cup is removed from the drink bottle and screwed tightly with the wooden tray, it turns into a normal drinking cup. The shape of the tray can be designed as a small cup or other shapes in colors to add pleasure when you drink.

花的故事
Storys of the flowers

Vases Made of Mineral Water Bottles
矿泉水瓶花瓶

【时间】：2008
【Time】：2008

【奖项】：2009 第十一届全国美展获奖提名
【Awards】：Nomination for the 11th National Artworks Exhibition 2009

利用矿泉水瓶口径统一为 28mm 标准尺寸，设计成圆柱形空心花瓶造型，将废弃的矿泉水瓶旋钮在花瓶内部构成一个简易实用的环保花瓶，可以将鲜花插在矿泉水瓶中。此设计制作简单且成本极低，具有很好的推广价值和环保意义，花瓶材质为铝合金材料。

A bottle of mineral water (in a uniform standard caliber-size of 28mm) is designed into a shape of cylinder hollow vase, and the discarded bottle of mineral water is screwed inside the vase so that a simple, pragmatic and environmentally-friendly vase comes into being. Fresh flowers can be inserted into the bottle of mineral water. This design involves a simple workmanship and an extremely low cost, so it is valuable in popularization and significant in environmental friendliness. The vase is made of aluminum alloy.

Bamboo-covered Bottle Vase

竹编矿泉水瓶罩花器

【时间】: 2012
【Time】: 2012

竹编的暖水瓶已渐渐成为旧时的记忆，以 1.5 升的矿泉水瓶作为内胆而编织成老式暖水瓶样式的矿泉水瓶花器，唤起我们对过往平静朴素的生活的回忆。竹编水瓶罩底部镂空，可以直接套在 1.5 升的矿泉水瓶上，简易便捷。

The bamboo-covered thermos bottle has gradually become a memory. The bamboo vase in the old-style thermos bottle shape with a 1.5L mineral water bottle as inside will arouse our memory of the innocent past. The bamboo cover is hollow cut on the bottom, which can be directly put onto the bottle.

Jardiniere and Vase

花架与花器

【时间】: 2012
【Time】: 2012

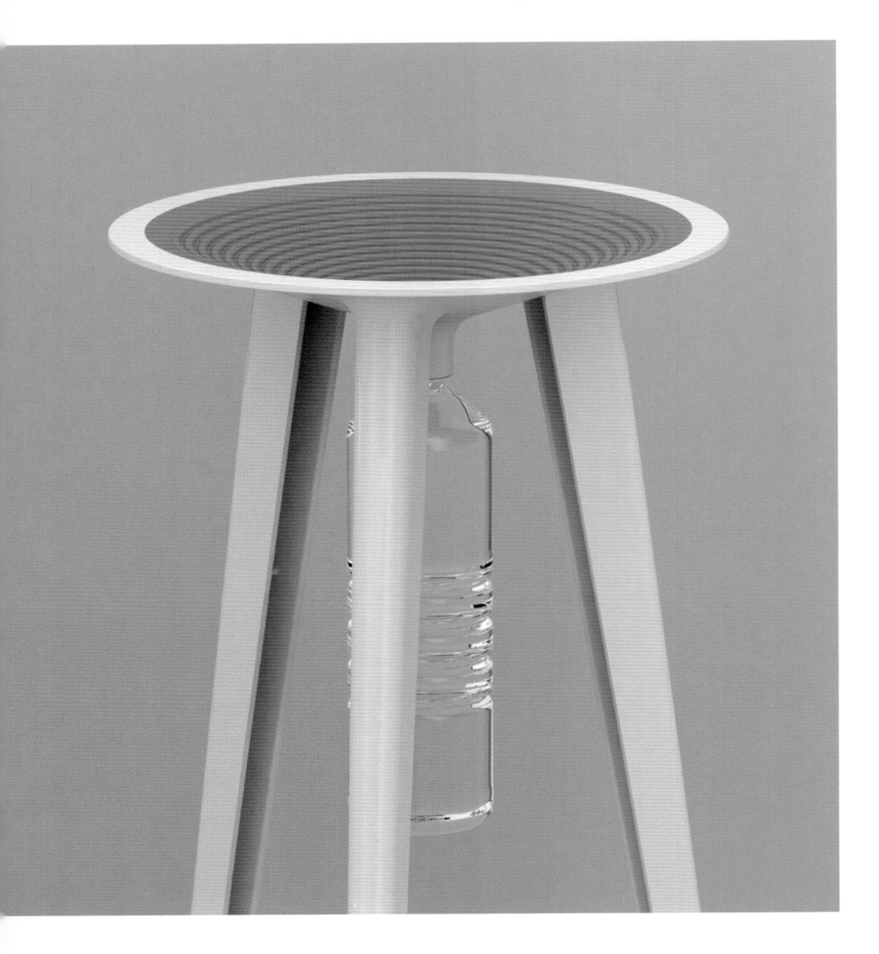

这是矿泉水瓶系列设计的另一件作品。养花的朋友们都会遇到这样的问题，我们时常因为浇花后，水会渗出花盆而烦恼。为此设计了这样一个花架，不但解决了浇花渗水的问题，同时也可以作为插花的花器使用。花架高650mm，花架托盘中央有漏水孔，托盘底部有螺纹可以将空矿泉水瓶旋转固定于底部（矿泉水瓶多为标准28mm 口径，可通用）。花架的托盘为漏斗形，内有环形波纹，适合固定各类大小的花盆。浇花时如果水渗出会沿漏斗形托盘流入矿泉水瓶中。不放花盆时，矿泉水瓶注入清水可作为插花的花器。整件作品的材质为塑料。

This is another design in the mineral water bottle series. People that like to grow plants all encounter such a problem that after watering the plants, the water will overflow from the flowerpot. Therefore, this jardiniere is designed. It not only will solve the water overflowing problem, but also can be used as a vase. It is 650mm high. On the center of the tray of the jardiniere there is a hole for water leakage, and on the bottom of it there is an screw to fix the empty bottle (the mineral water bottles mostly of the standard 28mm caliber, can be generally used). The tray is in a funnel shape with circular corrugations inside, suitable for fixing pots of different sizes. When people water the plants, if there is too much water, it will flow into the bottle along the funnel-shape tray. If no flowerpot is placed, the bottle can be filled with water and used as a vase. The material of the whole work is plastic.

Watering Pot Vase

洒水壶花瓶

【时间】: 2005
【Time】: 2005

【奖项】: 入选 2007 意大利 “ceramics for breakfast ” 国际设计比赛
【Awards】: Selected by “Ceramics for Breakfast” Design Competition and Contests Italy 2007

洒水壶是用来浇花的，既然这样，鲜花能否从洒水壶中开放呢？应该是可以的，这件设计作品用白瓷制成，我们将鲜花插在洒水壶中，洒水壶的洒水口与壶体相通。这样，每日在为鲜花换水的时候，我们无需把花一根根取出，只需将壶中的水从洒水口倒出，再一并将水从此处注入就可以了。

It is known that the watering pot is used for watering the flower. Such being the case, can the fresh flowers grow in the watering pot? There should be no problem. This design works will be made of white porcelain. We insert the fresh flowers in the watering pot, keeping the spraying mouth and the pot body interlinked. Thus, when we change the water for the fresh flowers everyday, it is unnecessary to take out the flowers one by one. The water can be spilled out from the spraying mouth, and then the fresh water poured in.

Violent Vase

暴力的花瓶

【时间】: 2011
【Time】: 2011

一把菜刀砍向一个斜圆形容器，并嵌在里面形成一个彰显“暴力”的个性花瓶。将菜刀劈砍的元素与花瓶配合不单单希望体现鲜明的个性，同时菜刀的握把成为花瓶的把手。这种不同器物形态相互组合的设计手法，考虑更多的应是组合后的功能与作用的担当。

The vase with a knife chopping at an oblique circular vessel becomes a “violent” one. The chopping of the knife not only shows the vivid character but also acts as the handle of the vase. Such combined design approach focuses more on function and effect.

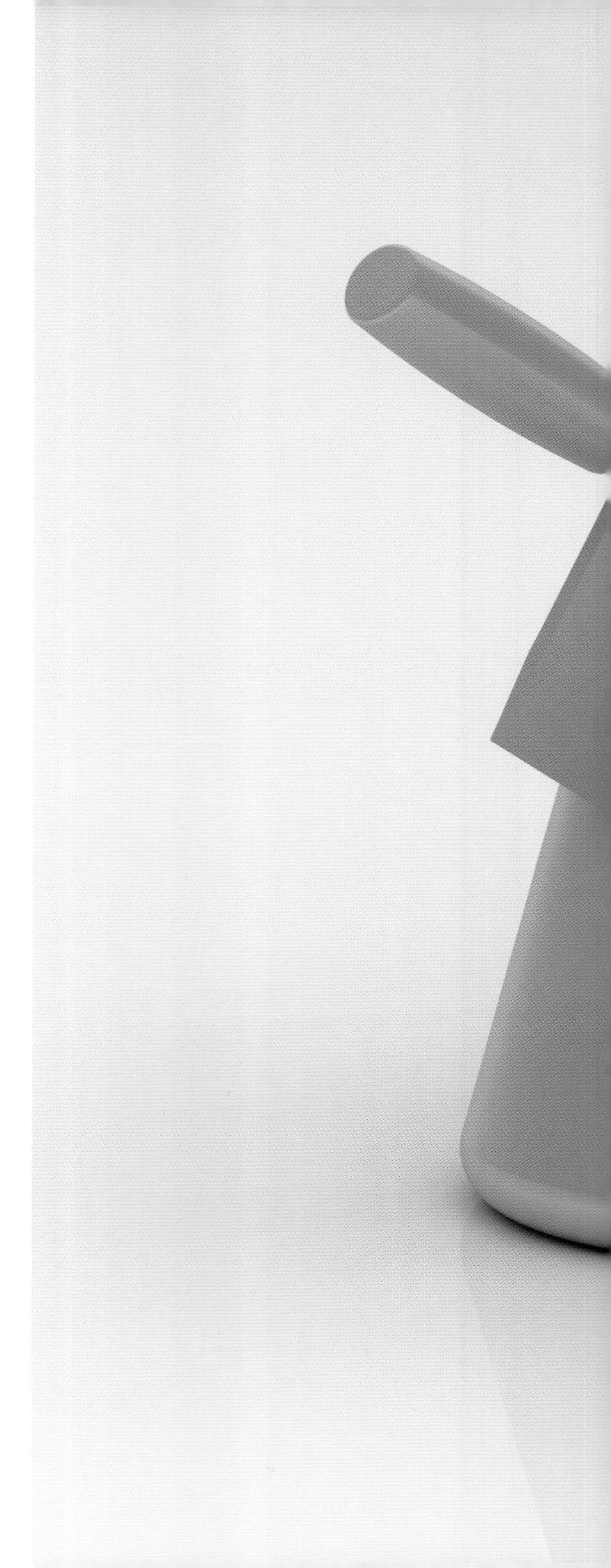

Ax-shape Vase

斧子花器

【时间】: 2009
【Time】: 2009

将五金店购买的铸铁斧子的一端切去成为一个可斜立的平面，换去木质斧柄改为塑料空心斧柄，便可成为插花的容器，以此体现破坏与重生这一对的矛盾概念。

An iron ax bought in a hardware store has been cut at one end to form a plane that can stand obliquely. The wooden ax handle is changed to a plastic hollow one so that it can be used to arrange flowers. This design shows the concept of conflict between damage and rebirth.

Grass–eating Monster

食草的怪兽

【时间】: 2009
【Time】: 2009

花盆里的情趣装饰品，植物没发芽的时候将这个张着大嘴的怪兽放置在花盆里面，当植物发芽向上生长时就会从大嘴里冒出来，好似怪兽在吞食植物一样。

This is a cute decoration in the flowerpot. Before the plant germinates, the monster with a big opening mouth is placed inside the pot. After the plant grows, it will come out from the mouth, as if the monster is swallowing it.

Morning Glory—Watering Funnel
喇叭花——浇花漏斗

【时间】: 2009
【Time】: 2009

喇叭花形的漏斗插入绿色盆栽的泥土中，不但可以作为花造型装饰，同时浇花时只要将水灌入花形漏斗中，漏洞底部均匀分布的漏水孔就可以起到浇灌花草的功能。材料为塑料，外部采用高光白色，内部用高光玫瑰红艳丽色，色彩明快且便于清洗。

The funnel in the shape of morning glory is inserted into the mud of the potted plant. It not only serves as a decoration but can be used for watering. The holes evenly distributed on the bottom of the funnel will water the plant. The material is plastic, polished in white outside and bright rose-red inside, easy for cleaning.

Flowerpot Mounting

花盆底托

【时间】：2010
【Time】：2010

这款花盆底托是为香港一家塑料家居企业设计的作品，在解决花盆底托基本功能的基础上强调构成感的装饰效果，整体镂空，绵延上升的圆环上端可以作为提手，便于搬运花盆。此作品尝试将花盆底托、装饰美化以及花盆搬运三者功效进行有机的结合。

This flowerpot mounting is a design from a plastic household company in Hong Kong. On the basis of solving the basic function of the flowerpot mounting, it also emphasizes the decorative effect. The entirely hollow cut and winding up ring can be used as a handle for conveying the pot. This work attempts to organically combine the functions of mounting, decoration and pot conveying.

工作的周边
Work ambience

Red Carp and Its Scales

【时间】: 2009
【Time】: 2009

【奖项】:
2009 美国 Sparkawards 星火国际设计大赛入选奖
入选 2010 中国包装技术协会设计委员会《中国设计年鉴》第七卷
【Awards】:
Selected Work of International Design Competition-Spark Awards U. S. 2009
Selected by Volume 7 of China Design Yearbook 2010, Design Committee of China Packaging Federation

这是一个挂在室内墙壁的硬币储藏器的设计。正面是两只红色的透明塑料材质的空心鲤鱼形状，它的内腔有容纳一个硬币厚度的空间，与红鲤鱼形状衔接的方形底面为铝合金材料。当硬币从红鱼的上部投入鱼的体内时，一个个硬币便组成鱼鳞的造型，要是取出这些硬币可以将储藏器取下，它的反面有两个螺纹开口，打开旋钮便可以方便地将硬币取出。

This is a design of a coin container suspended on an indoor wall. On its front side are two hollow shapes of red carp made of transparent plastics. In its inner cavity there is a space that can contain a coin. The square base linked with the shapes of red carp is made of aluminum alloy. When coins are cast one by one into the body of the red carps from their top, these coins will constitute shapes of fish-scales. To take these coins out, one can take down this container first. There are two spiral openings on its reverse. You can conveniently take the coins out after having unscrewed the knob.

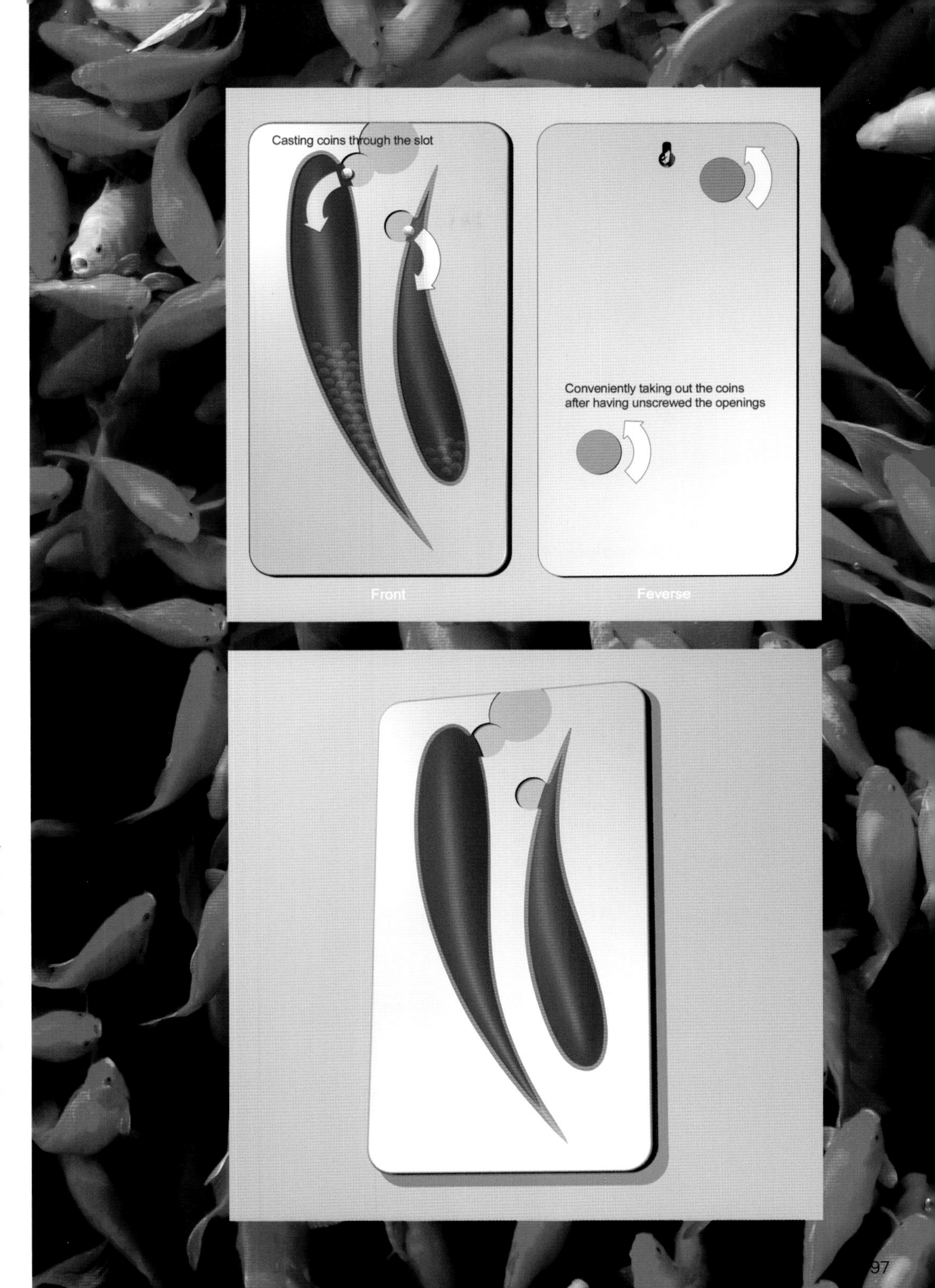

Man–Shaped Book Holders Series

小人造型系列书夹

【时间】：2007
【Time】：2007

【奖项】：入选 2008 中国包装技术协会设计委员会《中国设计年鉴》第六卷
【Awards】：Selected by Volume 6 of China Design Yearbook 2008, Design Committee of China Packaging Federation

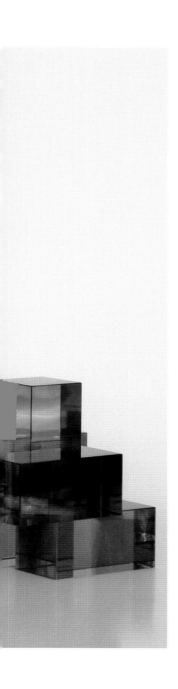

将琉璃或铝块材料加工成截面为 3×3cm、长 6cm 的立方体，每五个立方体组成各种姿势的小人形状，每个琉璃之间用专业胶粘合，铝块用焊接方式衔接。这样可制成各式各样的小人造型琉璃和铝块书夹。

3x3x6cm cubes are fabricated from colored glaze or aluminum, with every five of which assembled into man shapes in various poses. Professional glue is used to join colored glaze cubes, and welding is used to join aluminum ones. In this way, various man-shaped book holders of colored glaze and aluminum are made in different poses.

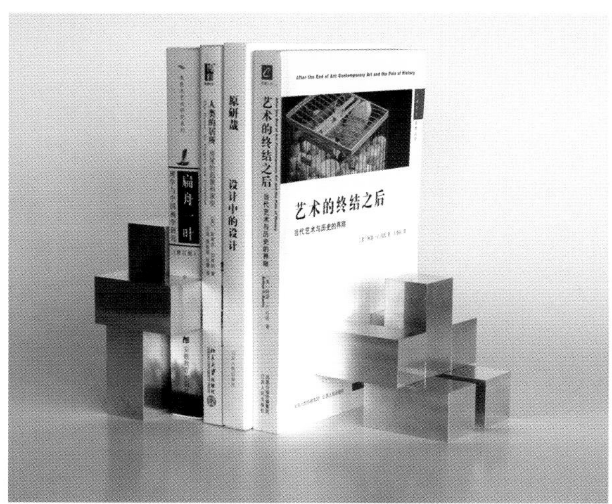

Series Book Ends in Vehicle Shapes

车造型系列书夹

【时间】：2007
【Time】：2007

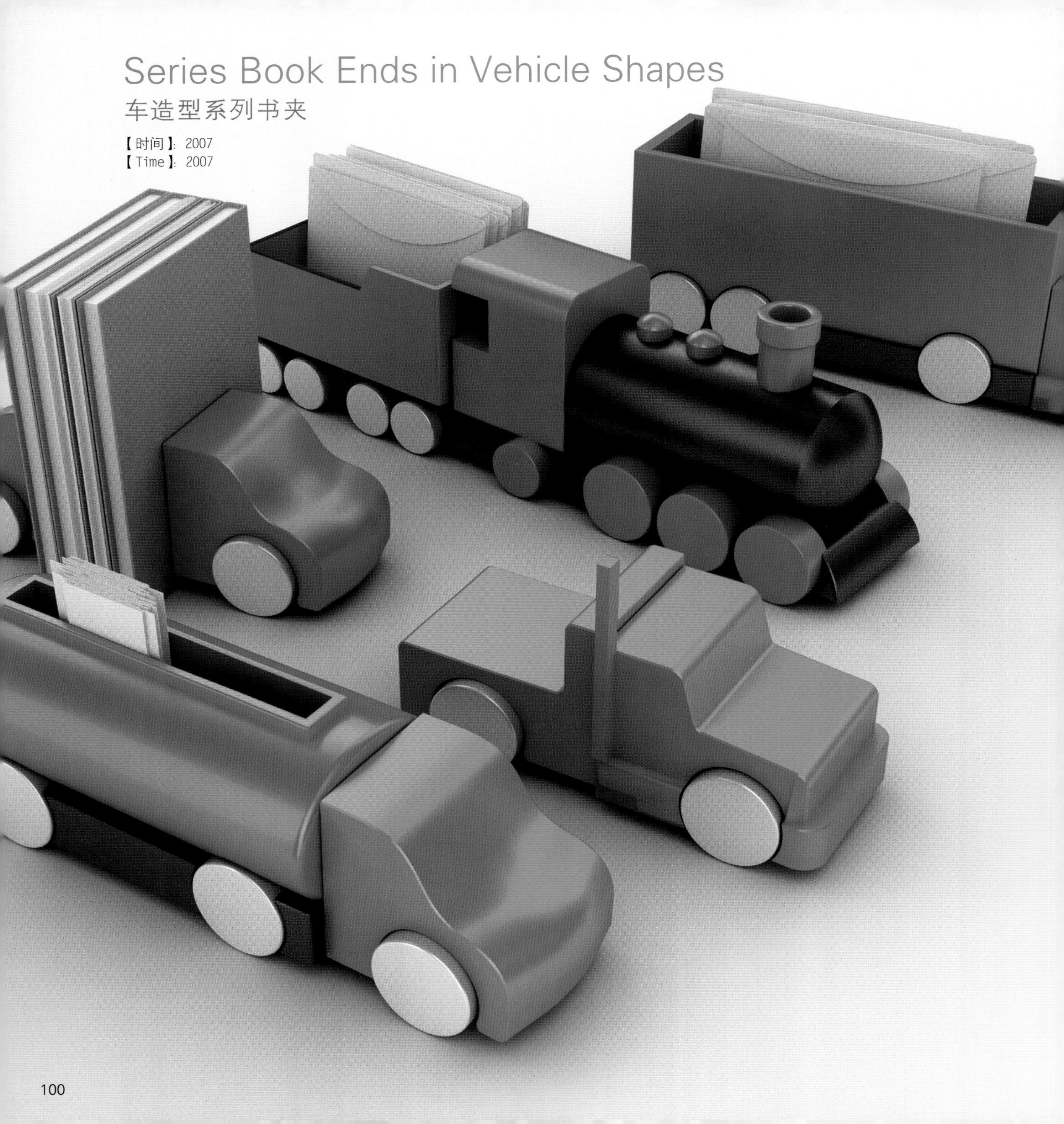

这个系列车造型书夹采用卡通夸张的造型，将车头和车身分开，利用车头和车身来作为书夹，同时每个车尺寸大小一致，这样各个车型可以随意互换车头和车身，带来不同的组合。大的车型如：卡车，火车，油罐车的车身同时可以作为放置名片，信封等杂物的储物盒。整体材料选用铝制，表面烤漆，白色与大红辅以黑色，对比艳丽。

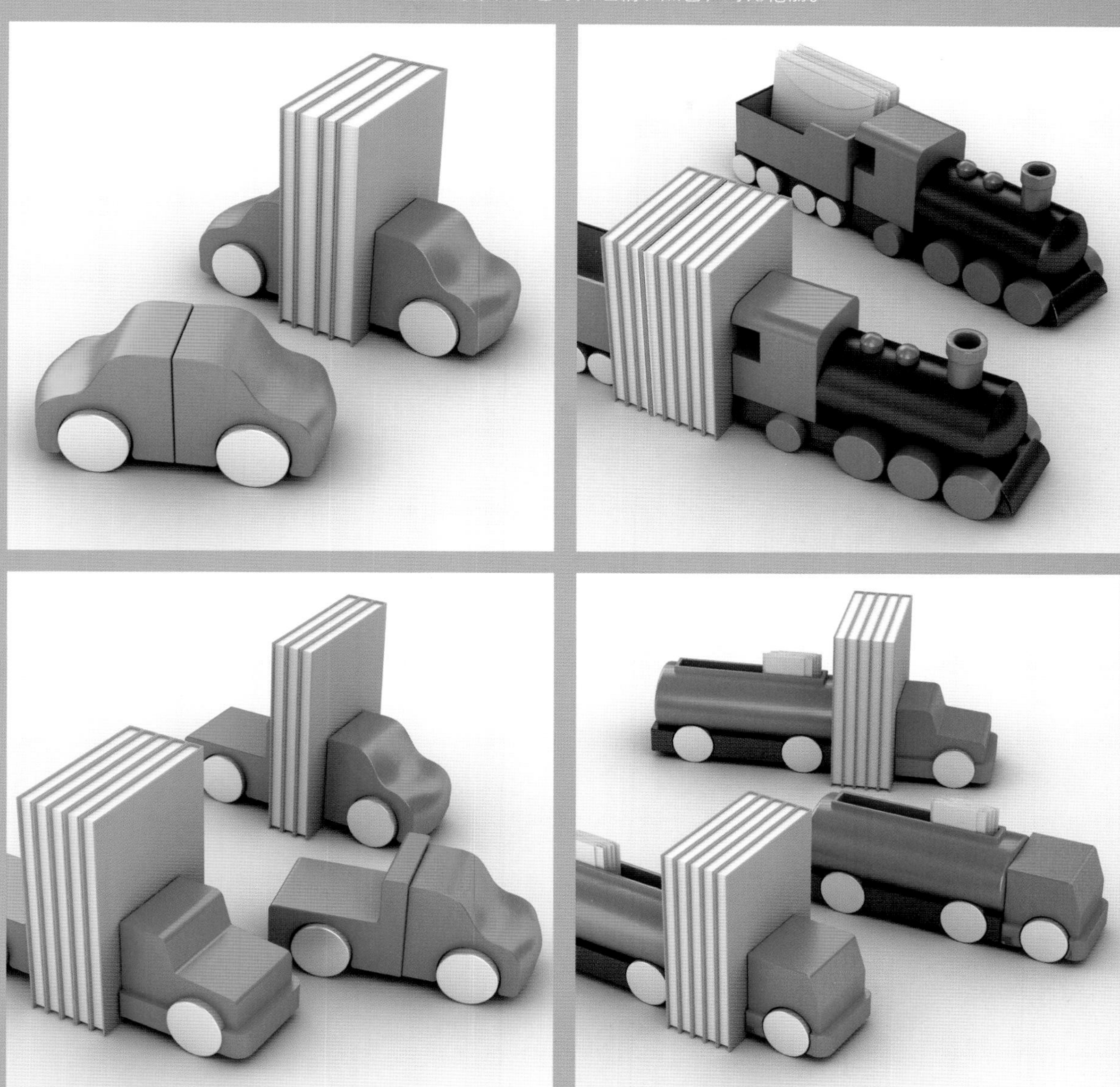

The series book ends use exaggerating vehicle cartoon shapes. Vehicle head and body are separated as the book ends. Every vehicle has the same size, so the heads and bodies can be interchanged to make different combination. Big vehicles such as trucks, trains and oil tanks, she bodies of which can also be used as storage boxes for name-cards and envelopes. The whole series are made of aluminum with baking varnish on the surface. The colors comprise white, red and black, in sharp contrast.

The Book with An Opening Door

开门的书

【时间】: 2011
【Time】: 2011

书籍装帧的概念设计。读一本书似开启一扇未知的门，虚掩的门缝里有一面镜子，透过镜子窥视到的不是书籍上的知识，而是重新看到了自己。

This is a conceptual design of a book. Reading a book is similar to opening an unknown door. There is a mirror in the seam of the unlatched door. What we see from the mirror is not the knowledge of the book, but ourselves.

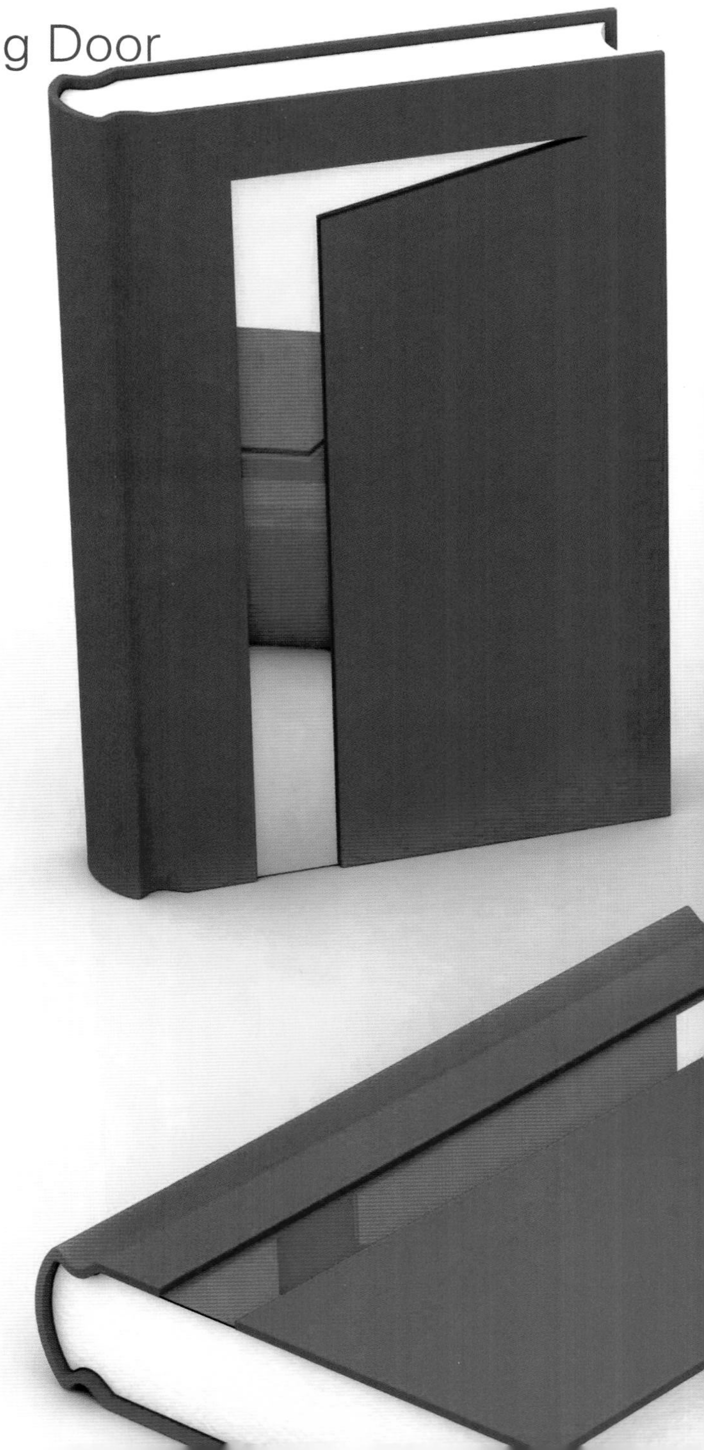

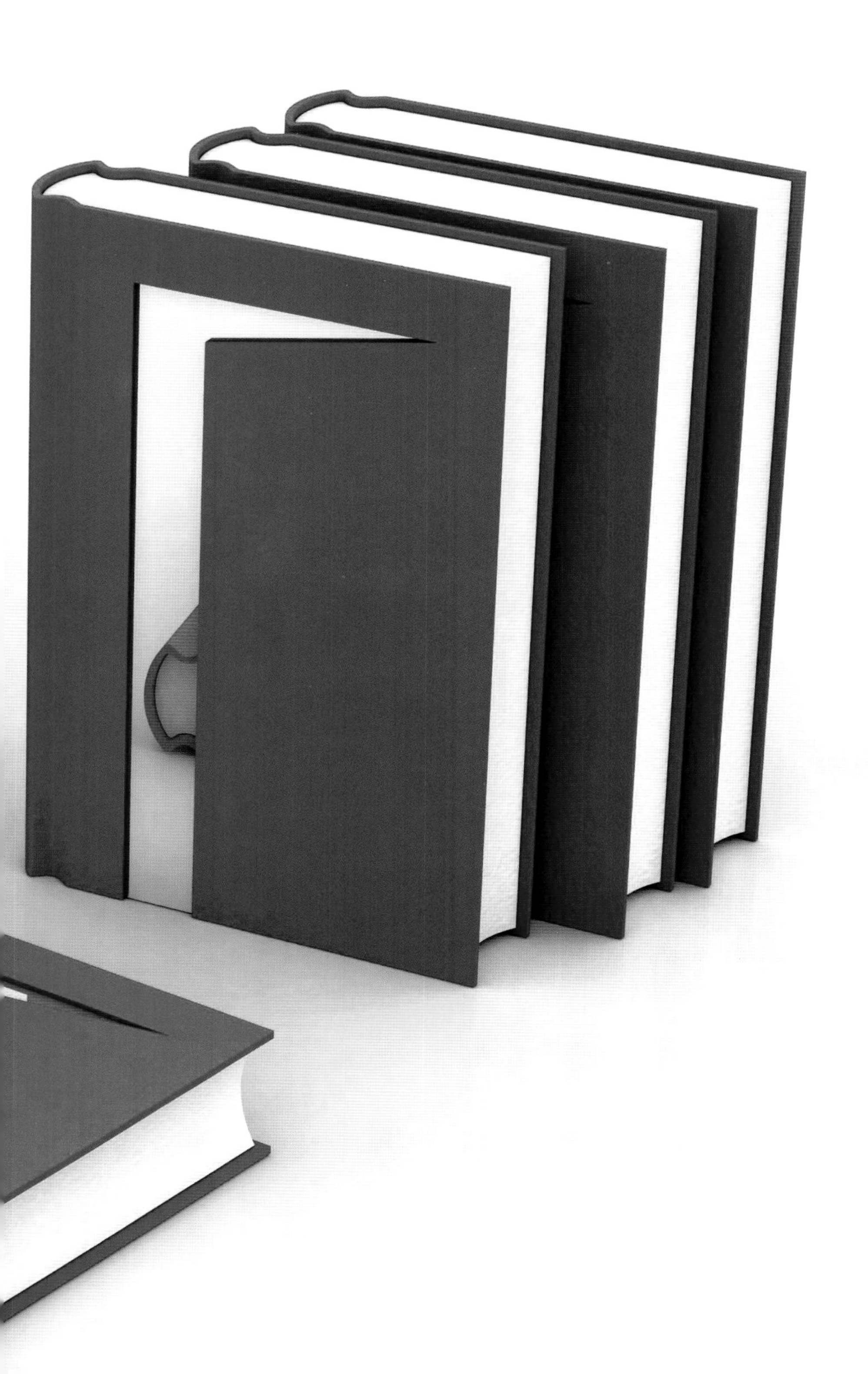

Work Time

工作时间

【合作】：于庆庆
【 Cooperated by 】：Yu Qingqing

【时间】：2008
【Time】：2008

【奖项】：
2008 韩国仁川国际设计比赛特选奖
入选 2010 中国包装技术协会设计委员会《中国设计年鉴》第七卷
【Awards】：
Special Selection of Incheon International Design Award 2008
Selected by Volume 7 of China Design Yearbook 2010, Design Committee of China Packaging Federation

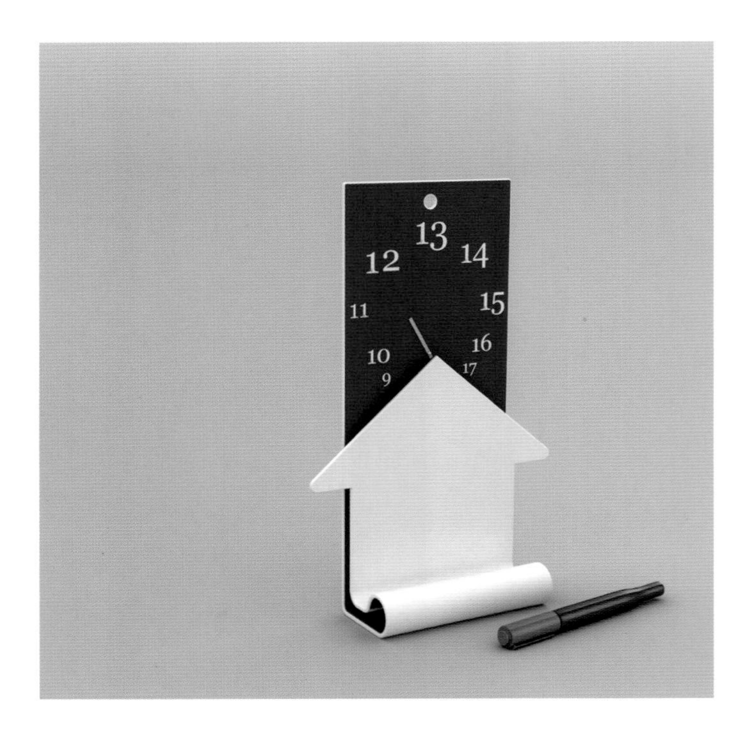

这是一个记录每日工作时间的计时器。计时器的时间刻度仅显示上班的时间，下班的休息时间隐藏在像家一样的白色房子后面。白色房屋形状的表面可以用记号笔写下当天的工作计划，擦拭后可以反复使用。
整件作品由一整块板材折压弯曲而成，钟表的机芯置于白色房屋造型与刻度板中间的夹层内。底部折压而成的圆筒型中间用于放置记号笔，这件作品可挂于墙上或置于桌面。

This is a timer that records daily work time. The scales of the timer only display the time at work, and hide the rest time off work behind a home-like white house. Marker pens can be used to write on the surface of the white-house-shaped board to note down today' s work plan, and the board can be used repeatedly after brushing.
The whole article is made by pressing and bending a whole board material. The clockwork is placed between the white-house design and the clock face. The folded cylindrical part at the bottom is used to contain the marker pens, and this article can be attached onto the wall or placed on the desk.

“Flower-fruit” Shape Drawing Pin Design

“花一果”图钉设计

【时间】2008
【Time】2008

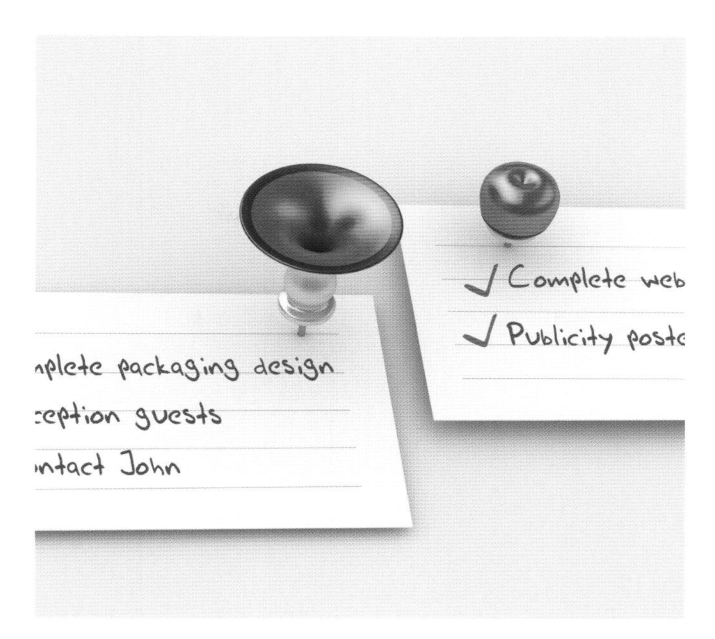

这是个由橡胶材料做成的图钉，将工作计划钉在墙上时，图钉像一朵朵盛开的小花。“花蕊”部分为实心橡胶，“花瓣”部分的橡胶材料柔软较薄，当工作计划完成时，用手按住花瓣向下弯曲，就能使花瓣反转形成一个果实的形状，表示着我们的工作有了结果。

Drawing pins are made of rubber material. When work plans are pinned on the wall, drawing pins look like small blossomed flowers. “Pistil” part is made of solid rubber and “petal” part is made of thin and soft rubber material. When a work plan is completed, press the petal and let it bend downwards to form a fruit shape which means our work has a good result.

Crash—Ornamental Magnet Design
坠机——装饰磁贴设计

【时间】: 2008
【Time】: 2008

【奖项】:
2008 意大利“Beyond Silver”国际设计比赛入选奖
入选 2010 中国包装技术协会设计委员会《中国设计年鉴》第七卷
【Awards】:
Selected Work of “Beyond Silver” International Design Competition Italy 2008
Selected by Volume 7 of China Design Yearbook 2010, Design Committee of China Packaging Federation

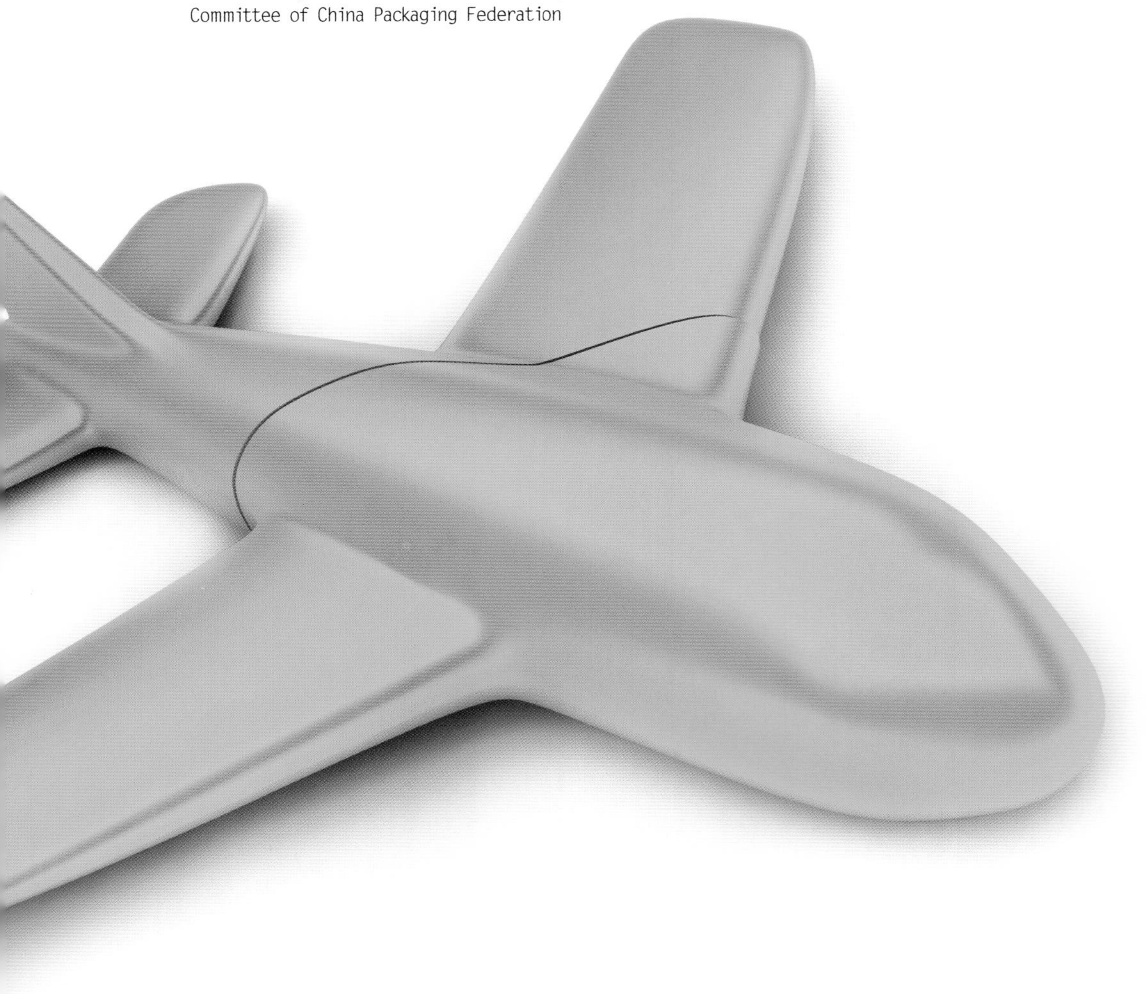

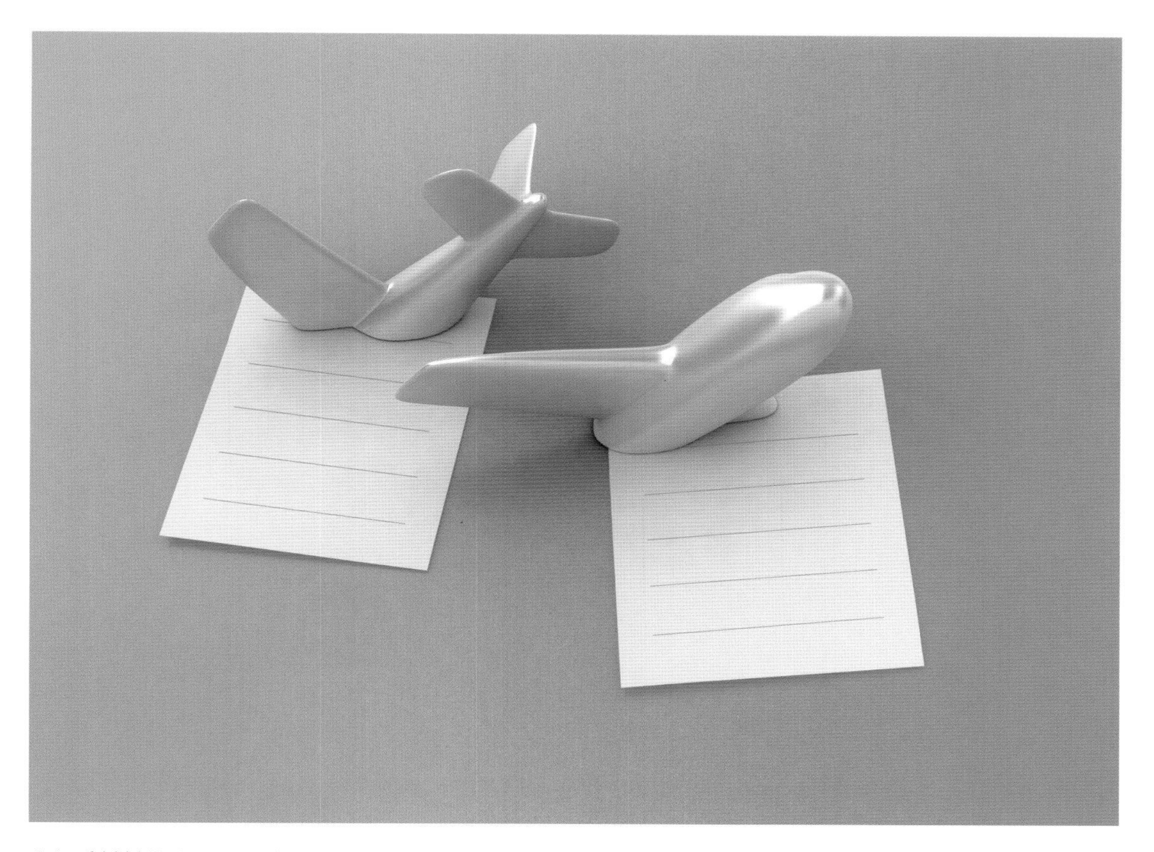

由银质材料做成的飞机造型，从中部一分为二，好似坠机的感觉。断开的截面嵌有磁铁，既可以相互衔接，两部分又可以吸附在铁板或冰箱之上。

This plane shape' s ornamental magnet design is made of silver and broken in the middle part,just like “a falling plane” . Magnet has been inlaid into the split sections. So the two parts can be either joined together or attached to a iron-panel or a refrigerator separately.

Record Clock
记录钟

【时间】: 2008
【Time】: 2008

可以将一天的工作计划写在便签纸上，用普通的塑料图钉定在钟的时间刻度旁作为提醒。圆形挂钟为分层结构，外部一层为标有刻度的塑料或铝合金表盘，表盘刻度旁均有圆孔。中间一层为密度板材质。使用时，将便签纸插入表盘与密度板两层相对应时间位置的缝隙间，用图钉透过表盘刻度旁的圆孔便可将便签纸钉在密度板上。

The schedule of the day can be written on a note fixed with a plastic thumb pin beside the time scale as a reminder. The round clock has several layers. The first layer outside is a plastic or aluminum alloy disc with scales, beside which there are holes. The intermediate layer is high-density board. A note can be inserted into the seam between the two layers at the position of the corresponding time. Fix the note on the high-density board from the holes beside the scales.

Tree-shape Key Hangers
树形钥匙挂钩

【时间】: 2008
【Time】: 2008

【奖项】: 入选 2010 中国包装技术协会设计委员会《中国设计年鉴》第七卷
【Awards】: Selected by Volume 7 of China Design Yearbook 2010, Design Committee of China Packaging Federation

铝合金材质的树造型，背面固定在墙上。众多延展的树枝可以悬挂钥匙类小杂物，既起到墙面装饰效果又具有收纳功能。

This is a aluminum alloy tree with its back fixed on the wall. The branches can be used to hang small things such as keys, with the functions of decorating the wall and collecting things.

Shelf Hook

书架挂钩

【时间】: 2011
【Time】: 2011

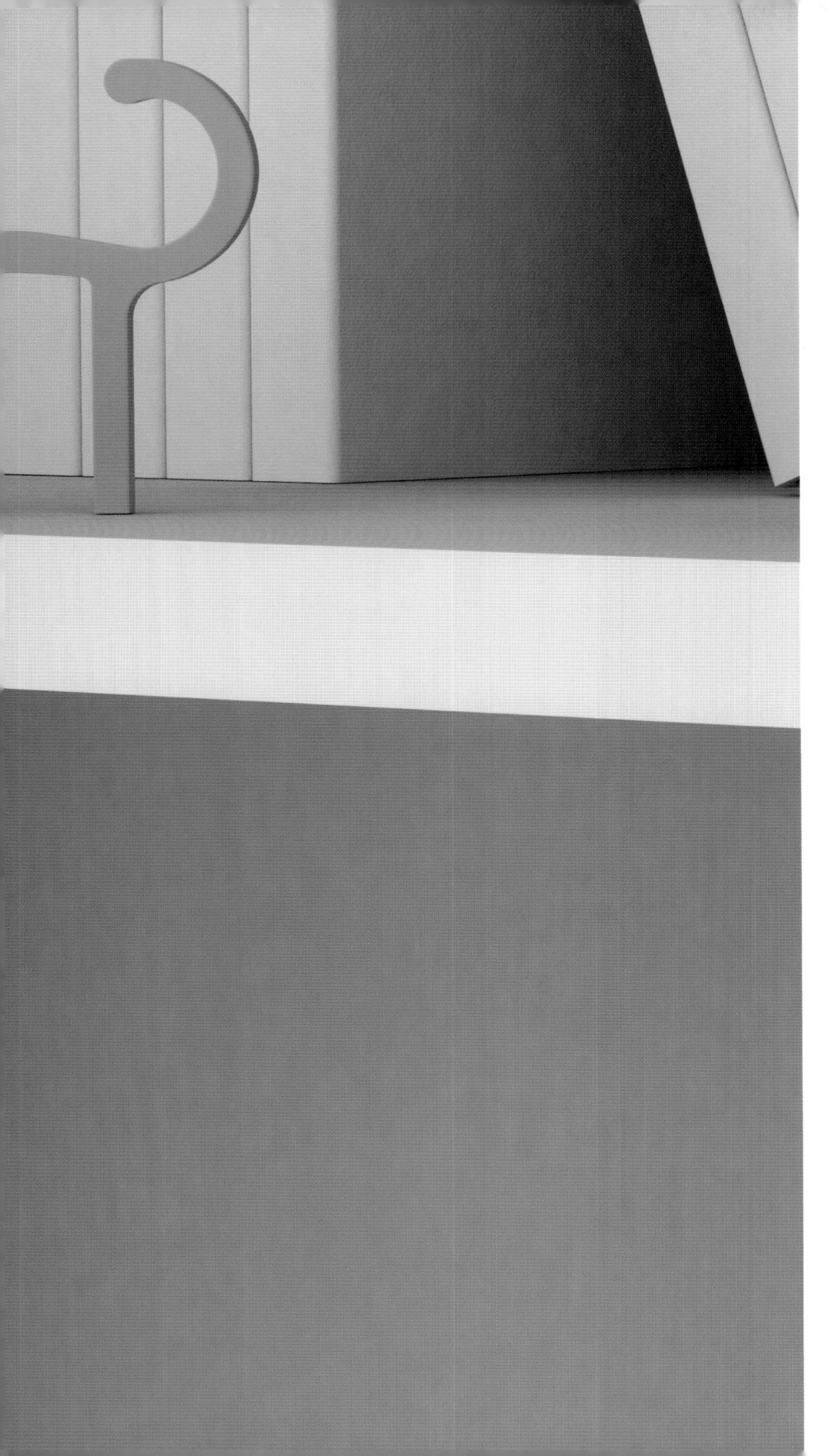

商场销售以及装修定制的书架多采用木工板材料，木工板加工成品后厚度大多为 20mm，利用这样一个标准尺寸，设计一款卡在书架面板上的挂钩。选用传统纹样的猛兽造型，挂钩材质为塑料，兽形挂钩两腿间距 21mm 可以卡在书架面板上，卷曲的兽尾可以悬挂钥匙以及背包等杂物。

Book shelves are usually made of wood boards, the thickness of which after processing is usually 20mm. By means of the characteristic of such a standard dimension, a hook clamped on the panel is designed. A traditional beast shape has been used, and the material is plastic. The space between the two legs of the hook is 21mm. It can be clamped onto the panel of the shelf, and the curling tail of the beast can be used to hang little things as keys and bags.

Touchable Title of Books

可以触摸的书名

【合作】: 杨子鹏
【 Cooperated by 】: Yang Zipeng
【时间】: 2012
【Time】: 2012

盲人书一排排放在书架上，盲人只有取下每一本书才能知道书名。为此我们在做书籍装帧时，在书籍边上做出可以触摸的盲文，方便盲人在书架上选择图书。
此设计制作较为简单，通过裁切的方式将书的每一页相对位置做出凸起的盲文，盲人通过触摸书边即可轻松辨别出书的名字。

Although there are rows of Braille books on the shelf, the blind have to take down every book to know the title of the book. Because of this, when we make a book , we make Braille on the edge od the book. Then it is convenient for the blind to select books from the shelf. To make this book is relatively simple, we can make the raised Braille on the relative position of each page of the book by measns of cutting. So it is easy for the blind to identify the name of the book by touching the edge of it.

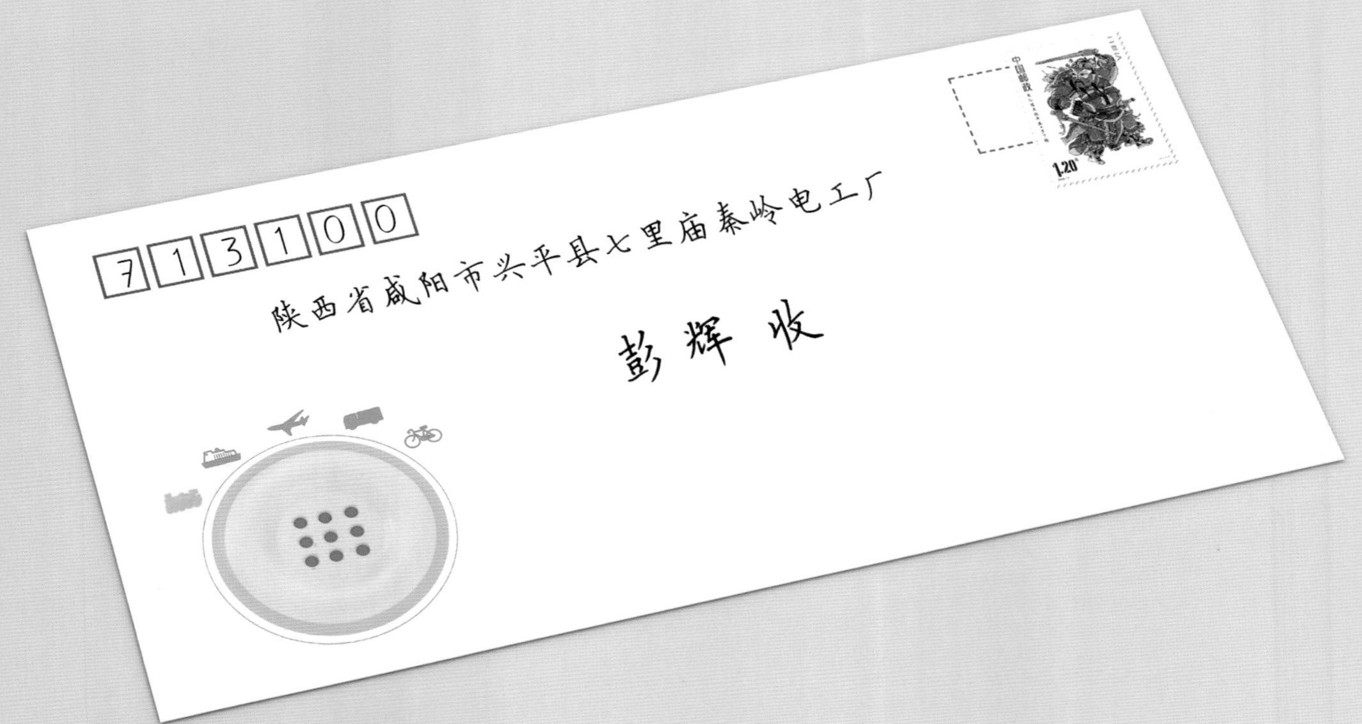

An Envelope Which Can Record the Sound During the Journey

可以记录旅途声音的信封

【时间】: 2009
【Time】: 2009

信封的正面有一个可以录播的装置，当我们将信封粘贴后，信封背面电源接通便可以录制旅途中各种交通工具的声音，它可以区分各种声音以便于录制一小段时间。当我们打开信封阅读信件时，播音装置启动，播放路途中各种交通工具的声音，同时作为区分交通工具声音的指示灯会亮起。

There is a recording and playing device on the recto of the envelope. When we paste the envelope, the power which is set on its reverse will be switched on to record the sound of various transportation tools. It also can distinguish different kinds of sound in order to record for short period. When we open the envelope to read the letter, the device will play the recorded sound. At the same time, the indicator lamp which is used to distinguish different kinds of transportation tools will be lighted.

Lego Modeling of Children's Phone
乐高积木造型儿童电话机

【时间】: 2011
【Time】: 2011
【合作】: 于庆庆
【Cooperated by】: Yu Qingqing

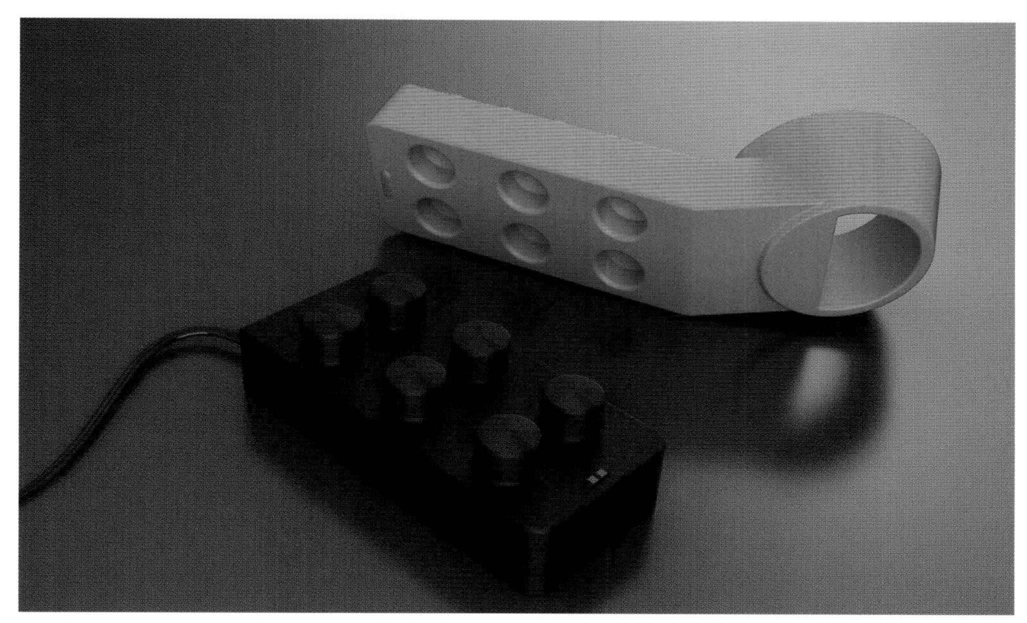

为一家电话机企业做的一款儿童分体电话机概念设计，采用了儿童喜爱的乐高积木造型元素。红色话机的按键凸起夸张，便于儿童操作，凸起的按键同时在造型处理上与黑色话机底座凸起的圆柱相互呼应，底座的凸起可以与话机背面圆孔一一对应，起到固定话机作用。听筒为镂空圆台形状，与整体方形产生对比，使得造型更为生动、富有变化。

This split telephone conceptual design for children adopts the elements of their favorite LEGO modeling, which is made for a telephone company.
The red phone keystroke is convex with exaggeration, which is convenient for children to operate. Meanwhile, raised keys in the modeling process echoes with the convex cylinder on the black telephone base, and the bulges on the base can match the back holes which will fix the telephone.
The receiver is a hollow frustum of a cone shape in contrast with the whole square which makes the modeling more vivid and changeable.

生活与概念

Life and concepts

The Tree Ring Web
年轮卷纸

【合作】: 马莲莲
【Cooperated by】: Ma Lianlian
【时间】: 2009
【Time】: 2009
【奖项】: 2009韩国仁川国际设计比赛银奖
【Awards】: Silver Prize of Incheon International Design Award 2009

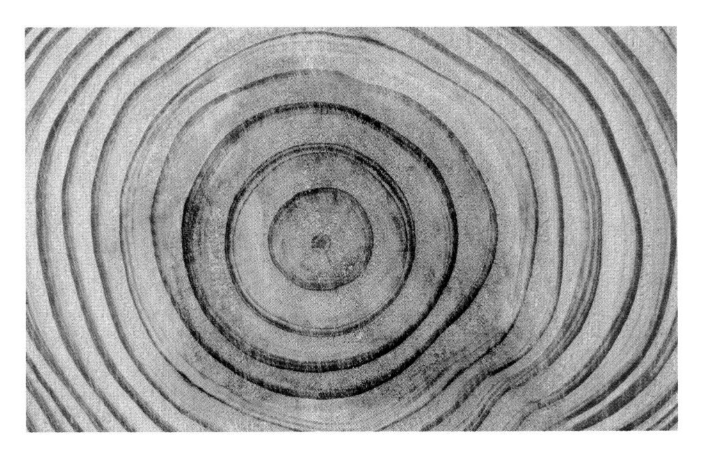

此设计的灵感来源于树木的年轮，旨在促进环保事业。材料上，呼吁人们节约用纸，珍惜资源；视觉上，向人们传递环保意识。设计没有增加原产品的成本但却赋予产品更深层次的内涵。

The inspiration of designing this web is from tree rings. And the web is aiming at promoting environmental protection. On the one hand, it calls on people to save paper. On the other hand, it transmits to people the awareness of environmental protection. The design does not increase the cost of the product but give the product a higher level connotation.

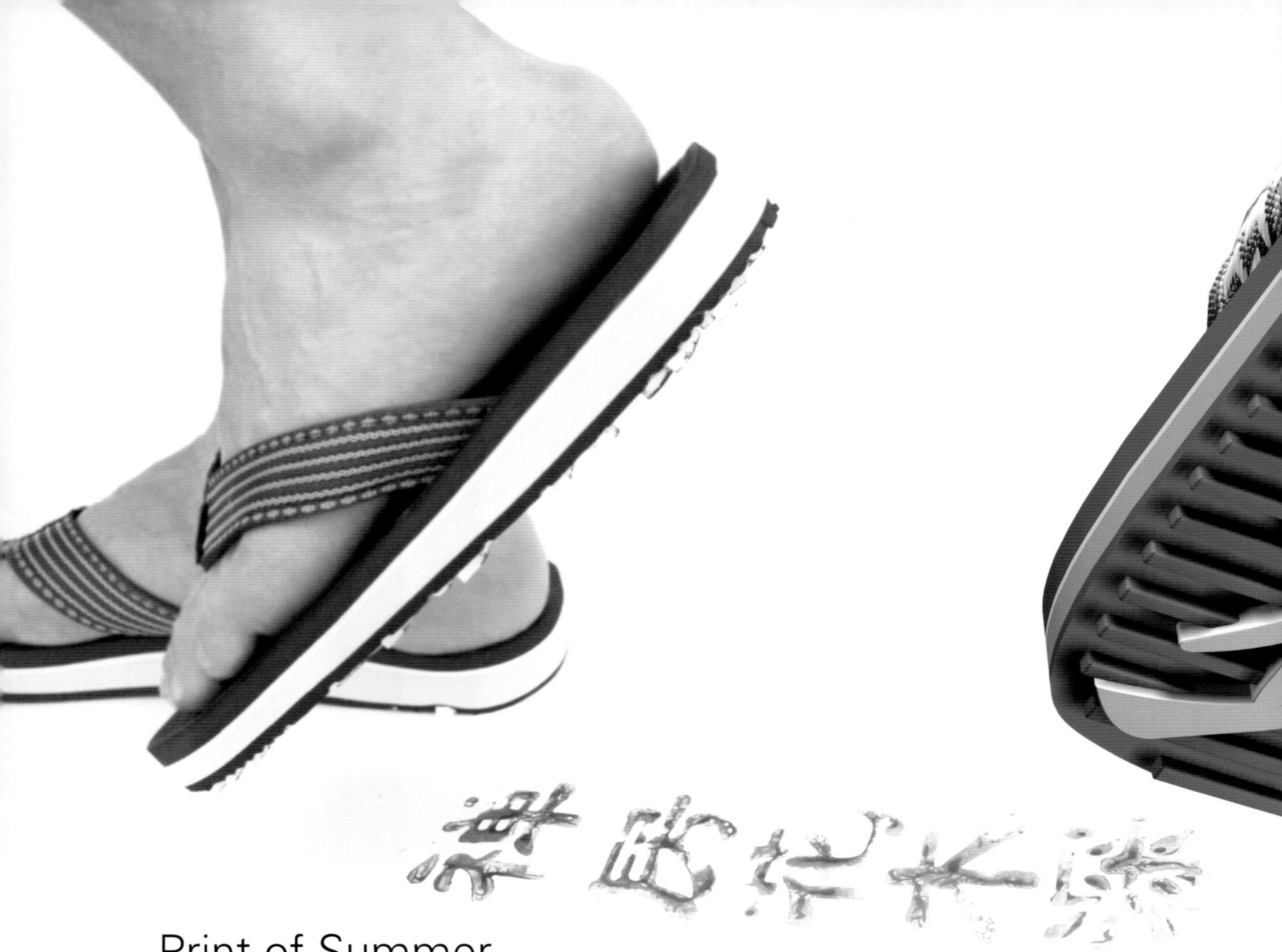

Print of Summer
夏日的印迹

【时间】: 2007
【Time】: 2007

【奖项】: 2007 北京 798 “旋转的硬币” 国际工业设计展优秀作品奖
【Awards】: Award of Excellent Work of SPIN COIN 798 International Industrial Design Expositions Beijing 2007

在沙滩鞋的鞋底附加凸起的文字，可以是你喜欢的诗句，也可以是你爱好的图腾，当你走过沙滩或是湿地，便会留下一串串诗句和图案。

Raised characters on the soles of slipper can be the poem or the totem you like. When you pass by a beach or a wet land, a poem or a pattern can be left behind...

Free Cage—Design of Bird Feeder

自由的鸟笼——喂鸟器设计

【时间】：2008
【Time】：2008

一款做成开放的鸟笼造型的喂鸟器，几根支架构成鸟笼的外形，小鸟可以自由穿梭于鸟笼内外。此设计旨在通过这样的造型表现“自由”与“束缚”这两个矛盾的概念，以设计的“善念”达到两者间矛盾的统一。

A bird feeder is made into an open cage shape. Several racks form the shape of the cage, but the birds can shuttle in and out of it freely. The design is to solve the two contradictory concepts of “freedom” and “restriction” by the idea “sympathy” .

Outdoor Flowerpot
户外花盆

【时间】: 2007
【Time】: 2007

将普通的花盆去掉底面，修剪成便于插入泥土的齿状。当我们在户外发现一株令人非常感兴趣的花时，可以用花盆将其圈起来。这样不会对花的原有生长环境产生影响，而更多地让引起人们对它的关注。

Remove the bottom from a normal flowerpot, and clip it into dentations so that it can be easily inserted into the soil. When we find a very interesting flower outdoors, we can enclose it with this kind of flowerpot. In this way, the original environment for the flower' s growth will not be affected. Moreover, the flower will arouse people' s attention.

Degradable Flowerpot Made of Straw Materials

可降解的秸秆材料花盆

【时间】: 2008
【Time】: 2008

秸秆材料主要指高梁杆、麦杆或枯树枝叶等自然植物的废弃物。发展中国家大多将秸秆作为燃烧物或动物饲料，这样既污染了环境，使用的效率也较低，这种可降解花盆是对秸秆材料再利用的一种尝试性的“绿色设计”。

将秸秆材料打碎成纤维状，与氮、磷、钾等肥料颗粒均匀混合后，用模具压制成花盆形状，花盆内侧喷涂可降解防水层，花盆的底托以精致的白瓷烧制而成，盆托可重复使用。在花盆内放置营养土，将植物的花籽埋于其中，待植物发芽长到适当程度时，连同花盆一起移栽到土壤中，花盆会随着自然界的降水而自然分解，同时，花盆在加工过程中掺入的各种肥料也会为植物提供所需养份。

本设计具有大量推广性和经济价值。

Straw materials mainly refer to wastes of natural plants such as broomcorn straw, wheat straw or branches and leaves of withered trees. In developing countries, straw is mostly used as combustion material or animal feed which could contaminate the environment and lead to low use efficiency. However, this kind of degradable flowerpot is a trial environment-friendly design of straw material reuse.
Break the straw materials into fiber form and uniformly mix it with fertilizer particles such as nitrogenous fertilizer, phosphorus fertilizer and potassium fertilizer. And then, use moulds to press it into a flowerpot and paint degradable water-proofing coating on the inner side of it. The bottom carrier of the flowerpot is made of exquisite ceramic white ware, which can be reused. Put nutrition soil into the flowerpot and bury flower seed into the soil. When the seed comes up and grows to an appropriate height, transplant it into the soil with the flowerpot. The flowerpot will be decomposed naturally due to rainwater. At the same time, various fertilizers mixed during the processing period of the flowerpot will provide the plant with required nutrition.
This design has a significant popularization capability and economical value.

Crossing Hook
十字架钉子挂钩

【时间】: 2011
【Time】: 2011

日常生活中常常见到将钉子钉在树上挂物品，针对这样的行为我设计了一个十字架造型的钉子挂钩，将对树的毁坏与挂钩的功能性矛盾地呈现。有时候，作为设计师，当我们无法改变大众日常习惯行为时，索性将其展现出来，取舍与态度就自然清晰了。

We sometimes see nails are fixed on trees to hang things. Therefore, aiming at such a behavior, we have designed a cross-shape nail hook to present the conflict between the damage on the tree and the function of the hook. As designers, if we cannot change people' s customary behaviors, we simply represent it to show the attitude.

One Tree, One Home—Design of Tree Guard Fence

一棵树一个家——树木护栏设计

【时间】：2010
【Time】：2010

当一粒种子从泥土中发芽希望能够长成大树的那一刻起，那片泥土就是它的家园。就像我们人类爱护自己神圣家园一样，我们没有任何权利剥夺属于树木的土地，这不仅仅出是于对树木这一生灵的敬重，更重要的是，一旦我们毁坏了森林，最终毁灭的是我们人类自己。因此做了这样的一件作品：将护树的栏杆做成房屋的造型，告诫那些砍伐者，这里是树的家园，树木与他们一样具有生命和感受。护栏的材料为白色塑料，钉在泥土里。

From the moment when a seed buds in the earth, hoping to grow into a big tree, that strip of land is its homeland. Just as we human beings love our own holy homeland, we have no right to deprive trees of the land belonging to them. It is not only from our respect for these living things of trees, and what is far more important is that we human beings will be ruined at last if we ruin the forest. So such works are produced: The fences guarding trees are made shaped like houses. They are to warn tree cutters that these places are the homeland of trees and trees have the same lives and same sense as people do. The fences are made of white plastic material and pegged into the earth.

3

Series Household Products Made by Discarded Bicycles

废弃自行车家居系列产品

【合作】：封冰，姚江，陈振译，贡琰，衡小东，刘蕊
【Cooperated by】：Feng Bing, Yao Jiang, Chen Zhenyi, Gong Yan, Heng Xiaodong and Liu Rui
【时间】：2006
【Time】：2006
【奖项】：
2006 中国家具设计大赛三等奖
入选 2008 中国包装技术协会设计委员会《中国设计年鉴》第六卷
【Awards】：
The Third Prize of China Furniture Design Contest 2006
Selected by Volume 6 of China Design Yearbook 2008, Design Committee of China Packaging Federation

这是配合 2006 年中央电视台毕业歌“放飞梦想”南京艺术学院专场演出中的“变废为宝”节目而设计的系列作品。系列作品所有选材均来自废弃自行车的零部件，依据部件的造型特征展开家居用品系列创意，并与南艺产品设计专业研究生合作完成。

This series of works were designed for the show “Changing wastes to treasures” in the “Flying the Dreams” performance of Nanjing University of the Arts in CCTV 2006. All the materials of the series were components from discarded bicycles. Ideas were given to household appliances according to the characteristics of shapes of the parts. The works were done in cooperation with the postgraduate students of the university majoring in product design.

1

1

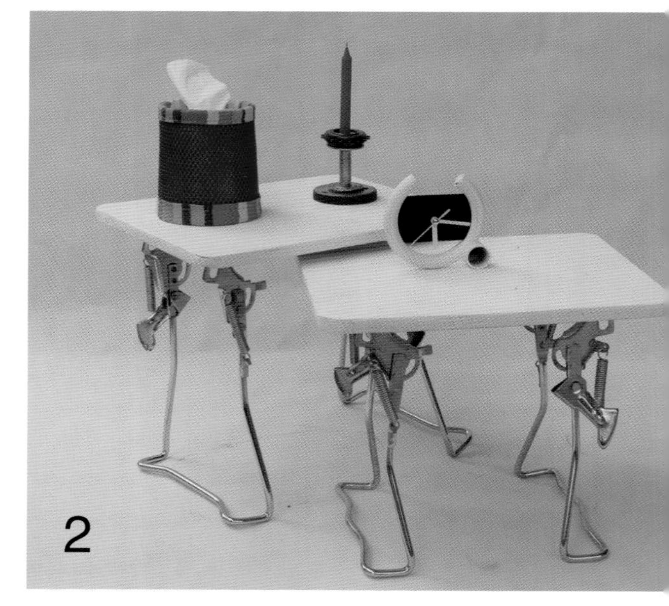
2

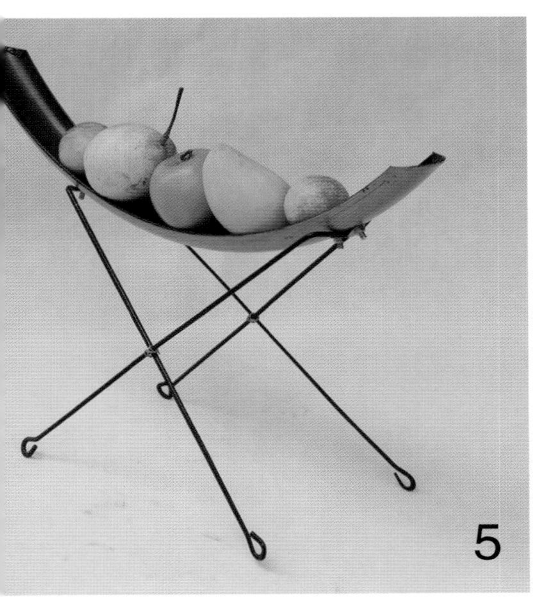
5

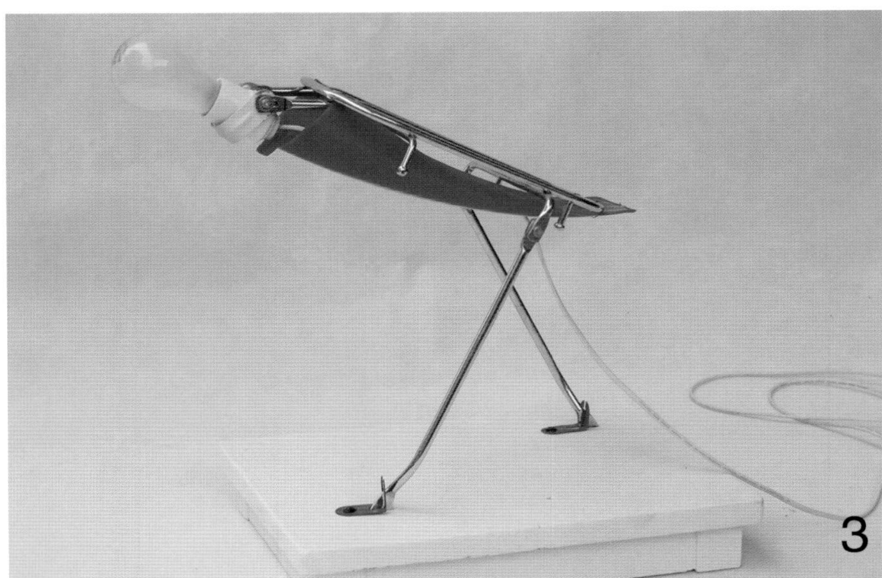
3

4

6

1、躺椅：自行车主要的支撑是中间的大梁，一般废弃车子主要的构架还是可以再利用的。平行放置，中间连接支撑，简洁的躺椅就诞生了。

1. Reclining chairs: The main support of a bicycle is the beam in the middle. Usually the main structure of a discarded bicycle can be reused by put horizontally with a connection between the structure. And then a simple chair comes to birth.

2、座椅：用停车柱和木板连接做了一对类似于小马扎的凳子，停车柱本身的曲线就极富美感，给产品增色不少。

2. Chair: the stop support of the bicycle and the wooden board are connected to make a stool. The curve of the stop itself already has its beauty.

3、灯具系列：利用废弃的挡泥板和货物架做了三款灯具。利用挡泥板作为灯罩，既有使光线间接照射的功能，又能和货物架巧妙结合，给人一种昂头前进的鸟的形象，具有趣味性。

3. Lamp series: three lamps are made of an abandoned flipper and a shelf. The flipper is used as the cover of the lamp to achieve indirect lighting. The smart combination of the flipper and tle shelf makes it like a striding bird with a raised head and has a lot of fun.

4、烛台系列：废弃的车铃铛和车轴承也可以再利用，做成烛台，增加了家的浪漫和温馨的气氛。

4. Candle holder series: discarded bells and bearings can also be reused to make candle holders, adding cozy atmosphere to home.

5、水果托架：利用自行车车轮铁制挡板做成的水果托架，美观实用。

5. Fruit stand: the iron baffle of bicycle wheels are used to make fruit stand, which is beautiful and useful.

6、挂钟系列：利用自行车上大大小小的飞轮、轴承还有链条，制作了三款有机械美感的钟表。制作简单，可以利用齿轮自身的美感发挥想象设计出不同的造型。

6.Series of wall clocks: Three clocks with mechanical beauty can be produced by different size of bike flywheels, bearings and chains. Making is simple. We can also design different shapes by using different gears.

Pet's Toy 宠物的玩具

【时间】: 2005
【Time】: 2005

【奖项】: 2007 北京 798 "旋转的硬币" 国际工业设计展优秀作品奖
【Awards】: Award of Excellent Work of SPIN COIN 798 International Industrial Design Expositions Beijing 2007

2005 年夏天，在南京金盛百货的五金店看到有塑料交通警示柱和成捆的麻绳出售，于是就在小店现场制作了一个猫扒爬架，将麻绳一圈圈紧紧地缠绕在警示柱上，收口处用细铁丝捆绑扎紧。带回家后妮妮、Poly、棉球都非常喜欢，每次它们睡醒了都争抢着用它作扒爪子功课。回忆起这件作品时我时常感叹：设计不是一个职业，是生活。

I saw the plastic traffic warning sign and linen ropes being sold in Nanjing Jinsheng Store in the summer of 2005. I made a small climbing stand for the cats in the store by winding the ropes tightly onto the warning sign, and binding off tightly with a fine iron thread. After I took it home, my cats loved it a lot, playing with it over and over after they woke up everyday. This makes me realize that design is not a job but a life.

Cable Coiled Dog

电线缠绕的小狗

【作者】: 路益（外甥）
【Author】: Lu Yi (my nephew)

【时间】: 小学三年级 (20**)
【Time】: Grade 3 in elementary school

外甥送给舅舅的一件自己设计制作的作品，用电线缠绕的小狗。我很是喜欢，形态与结构，以及电线颜色的选择都恰到好处。

This is my nephew' s design, a cable coiled dog. I like its shape and structure as well as the colors he selected.

When the winter comes and the lake ices over, fishes would die from oxygen deficiency, especially in high-altitude areas. Many fishermen will regularly chisel holes on the ice to save the fishes.This oxygen-adding bucket can be placed in the lake before the icing up. This double-layered column floats with attached weight at bottom. The ice reaching 10 cm thick allows people to walk upon. Under this circumstance, the inner bucket filled with ice can be twisted out, thus letting in oxygen to the water. Meanwhile, the hole facilitates fish feeding and even fishing.

Fish Hole —The Oxygen–supplying Bucket

鱼洞——冬季鱼塘增氧桶

【合作】: 杨万里
【 Cooperated by 】: Yang Wanli
【时间】: 2012
【Time】: 2012

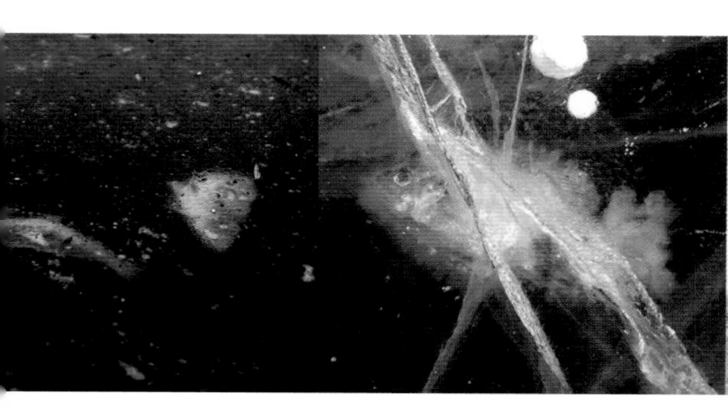

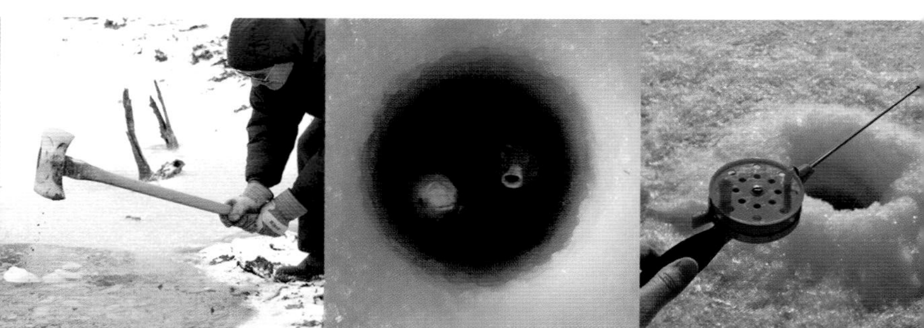

高纬度地区的鱼塘湖面冬季结冰，导致大量的鱼类因为冰层封水面而缺氧致死，因此许多渔人都会定期在封冻的湖面凿出许多通气孔来解决鱼类缺氧这一问题。为此我设计了一款“鱼洞”增氧桶，未结冰之前将增氧桶均匀投掷在湖里。桶为双层圆桶造型，底部有配重可使得桶体悬浮在水中。

当湖面结冰超过 10 厘米时，人可以安全在冰面行走，这时只需扭动并取出内桶，冰柱将随内桶一起取出，露出水面即可起到增氧的作用，同时也方便给鱼喂食或在冰面垂钓。

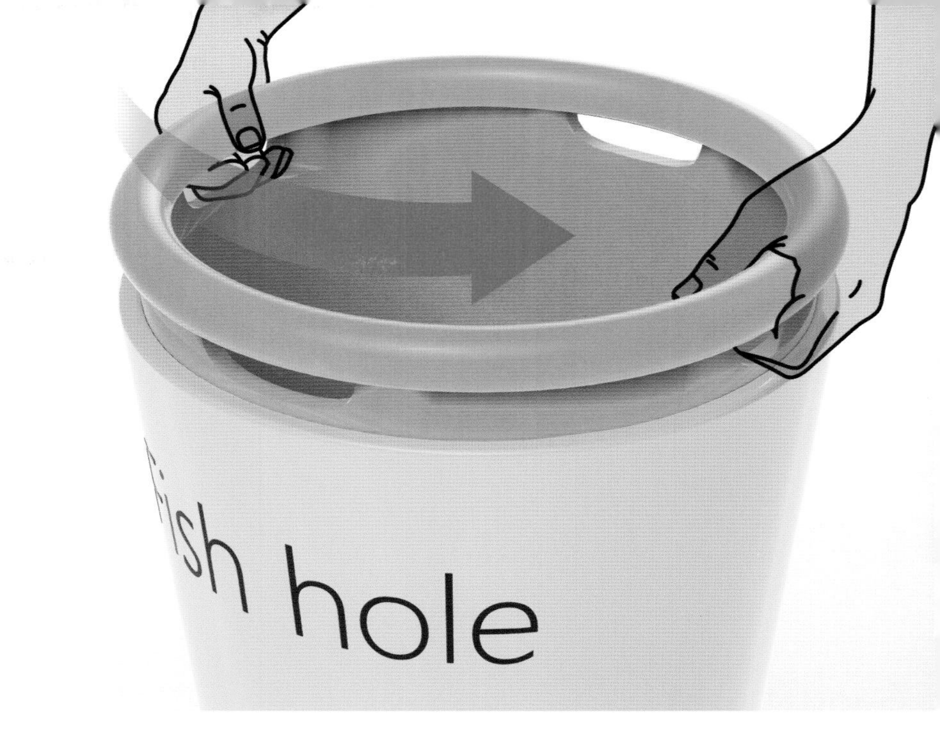
Fish hole

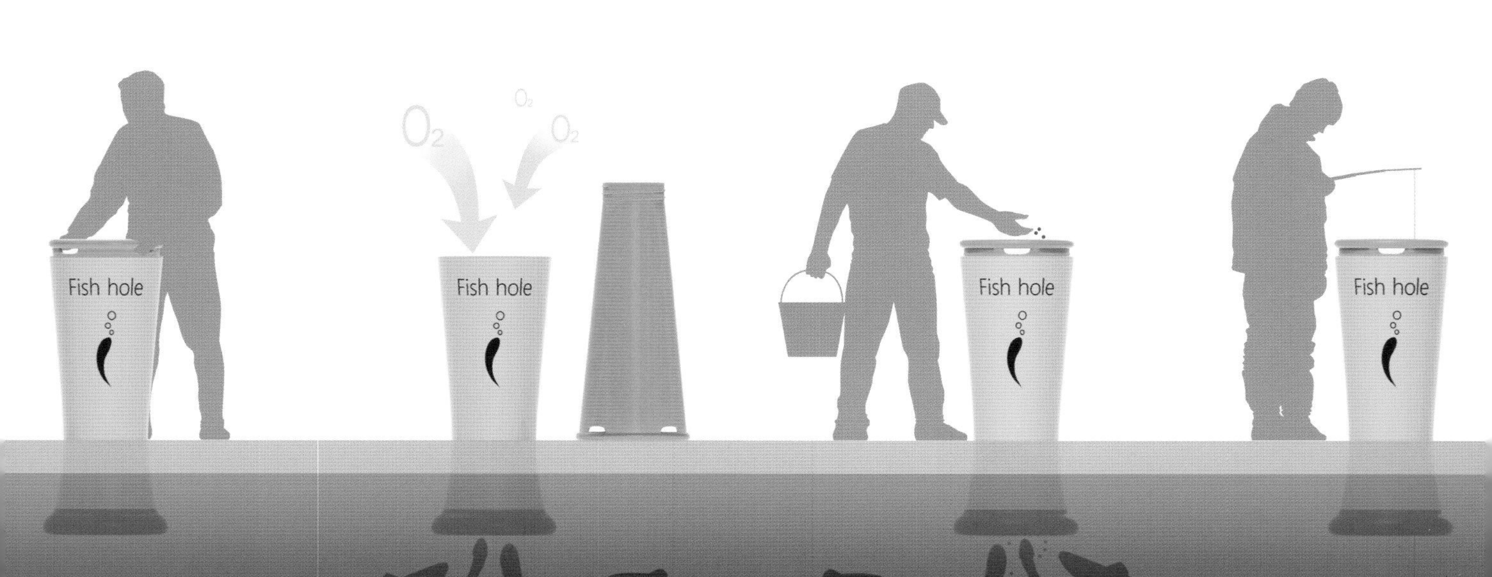
O_2
O_2
O_2
Fish hole
Fish hole
Fish hole
Fish hole

Series of Cartoon Models

系列卡通造型

【时间】: 2005-2007
【Time】: 2005-2007

【奖项】:
入选 2008 中国包装技术协会设计委员会《中国设计年鉴》第六卷
【Award】:
Selected by Volume 6 of China Design Yearbook 2008, Design Committee of China Packaging Federation

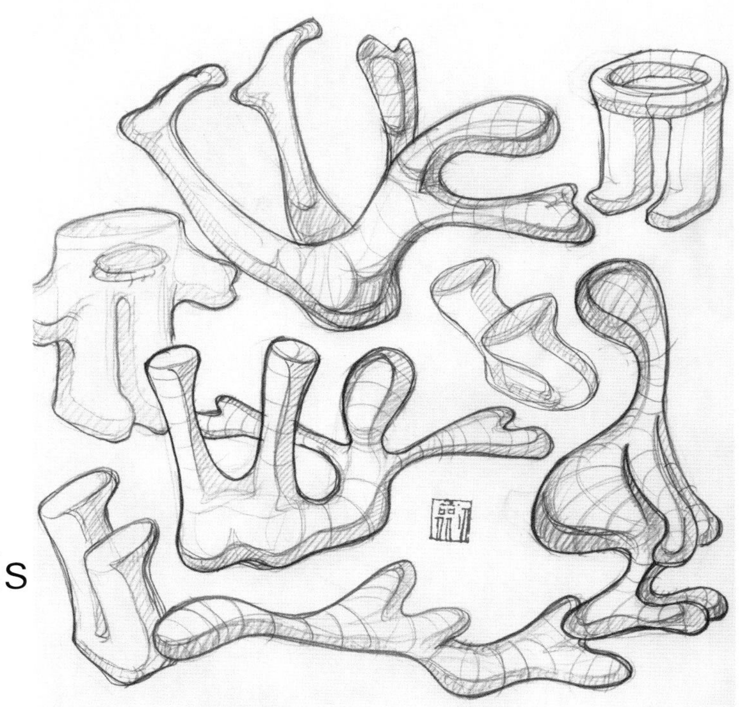

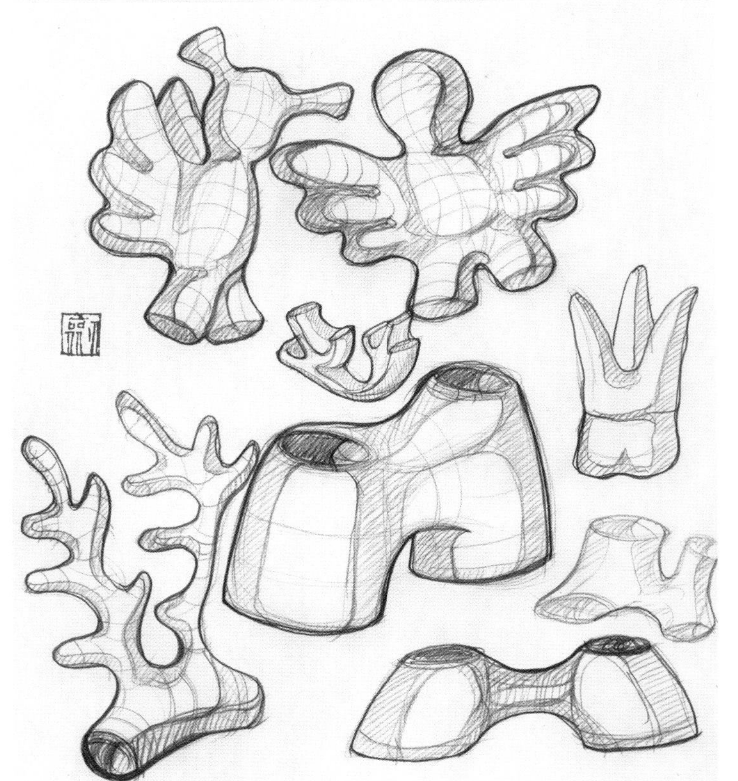

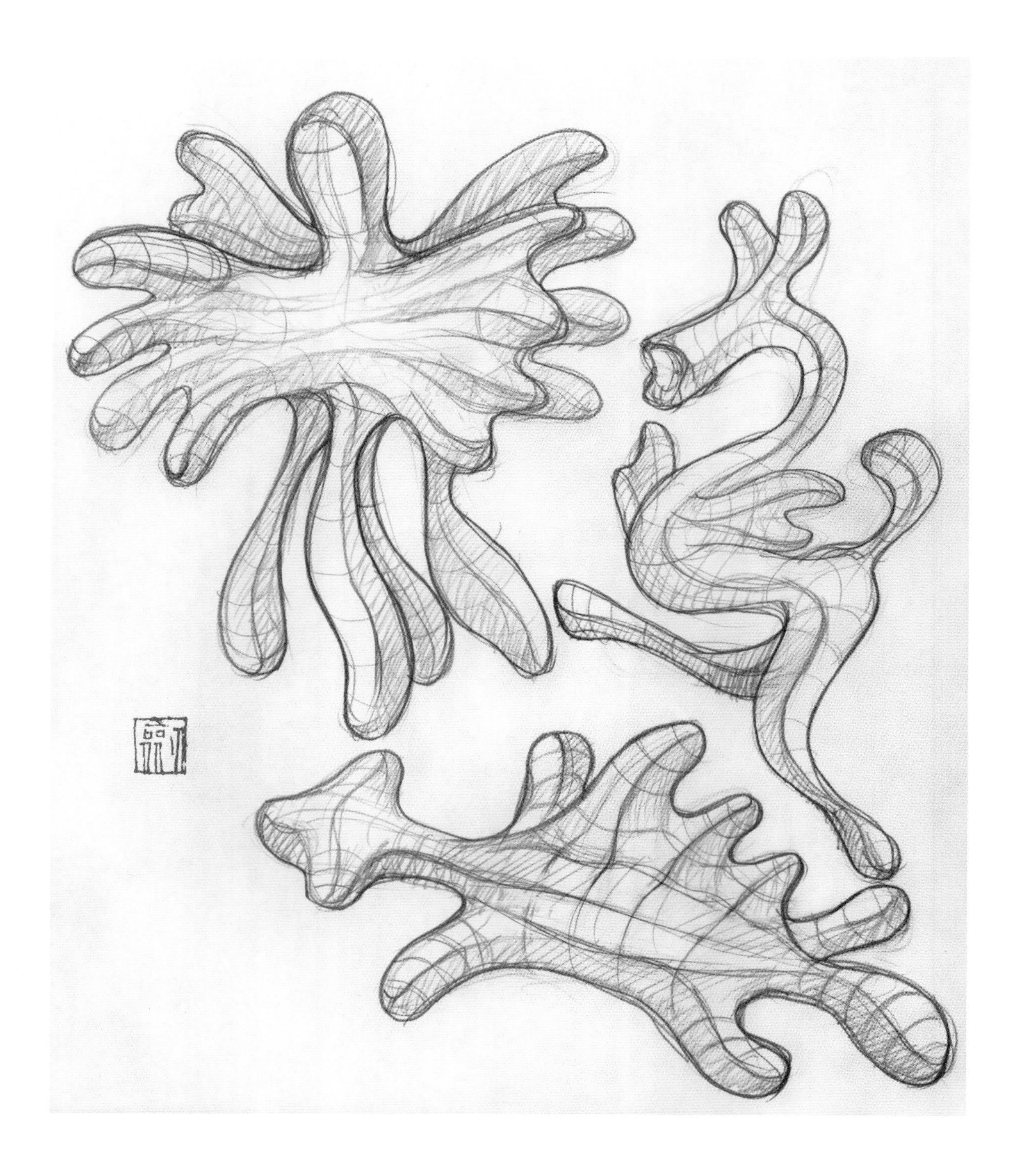

断断续续在两年间完成的系列卡通动物造型设计，这些造型有些根据现有动物造型夸张提炼，有些则是以形式感为原则的臆想造型。做这些卡通动物造型一方面是进行有机形态的练习，另一方面也希望自己三维软件不生疏。

Series of cartoon animal models have been completed during two years discontinuously. Some of these models are extracted from the animals' images by exaggerating, and some are created. The purpose of making these cartoons is to practice the shapes, meanwhile maintain my ability of using 3D software.

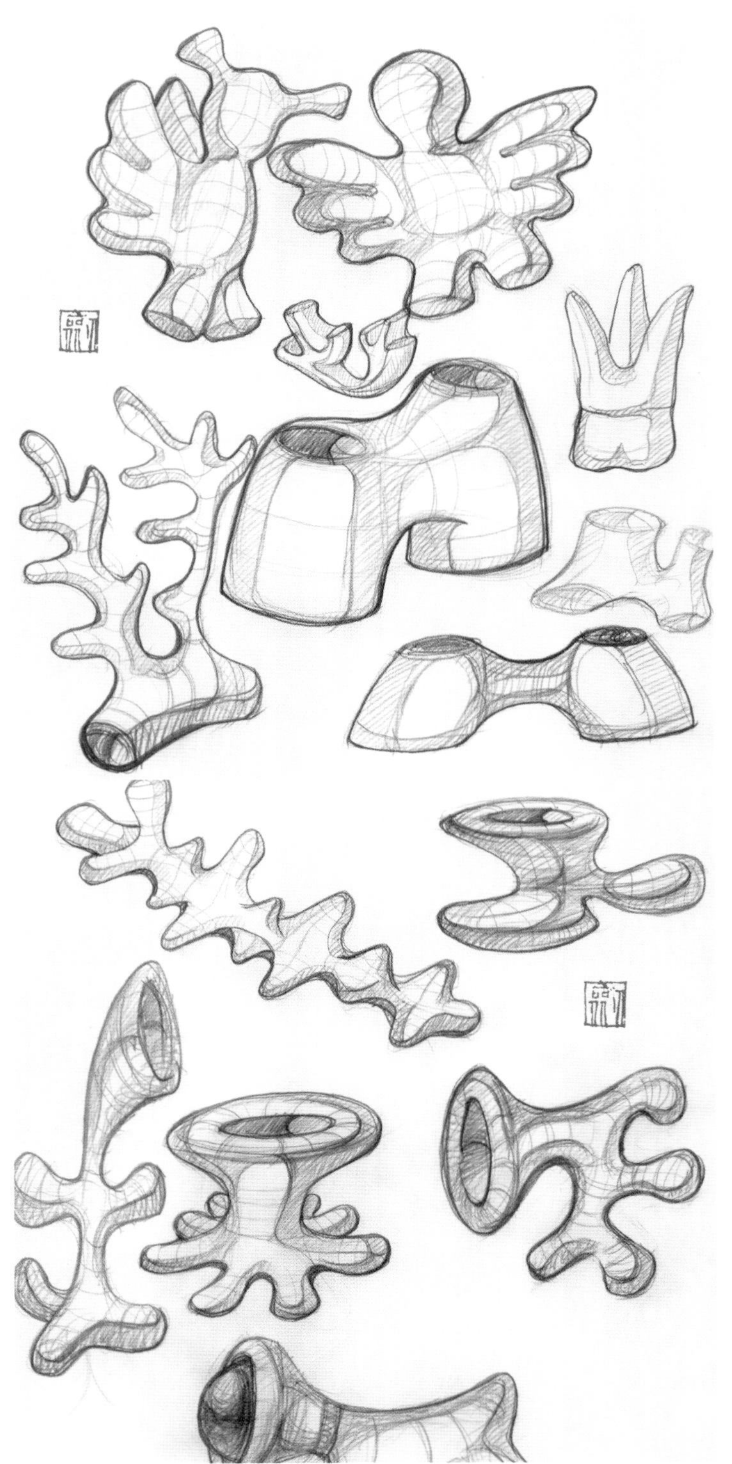

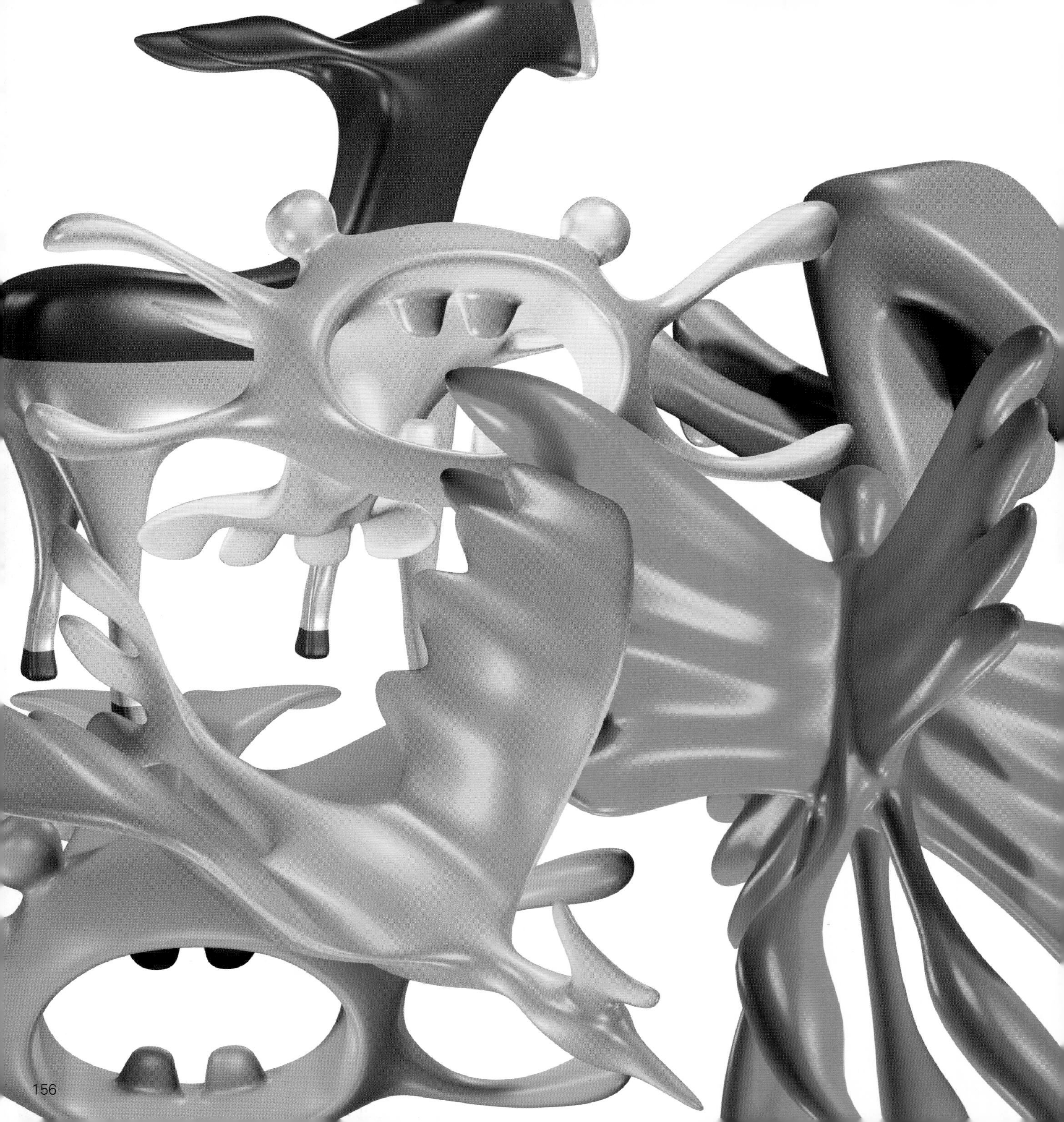

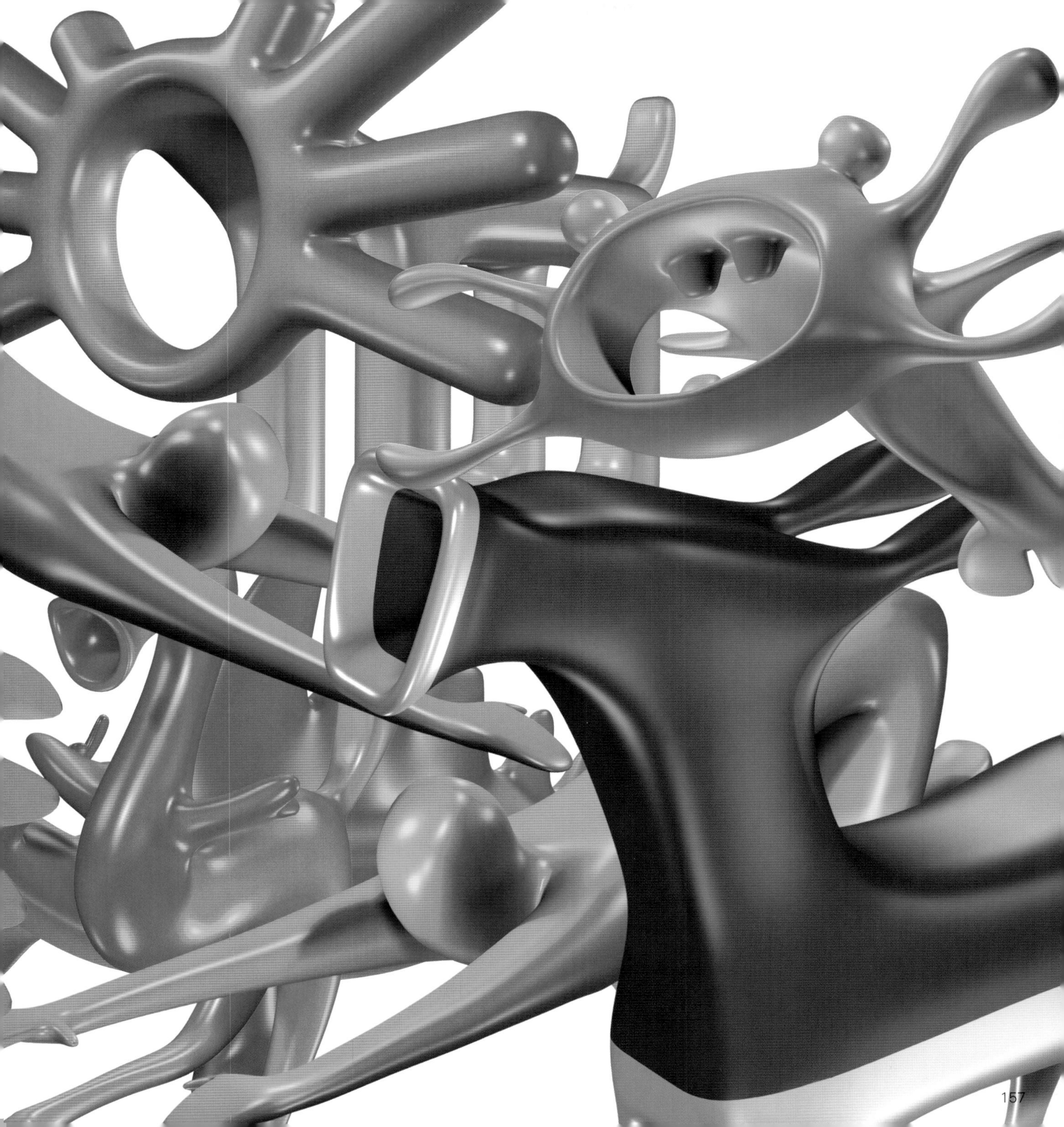

Public Facilities Series in Beijing Olympic Park
北京奥林匹克公园公共设施系列

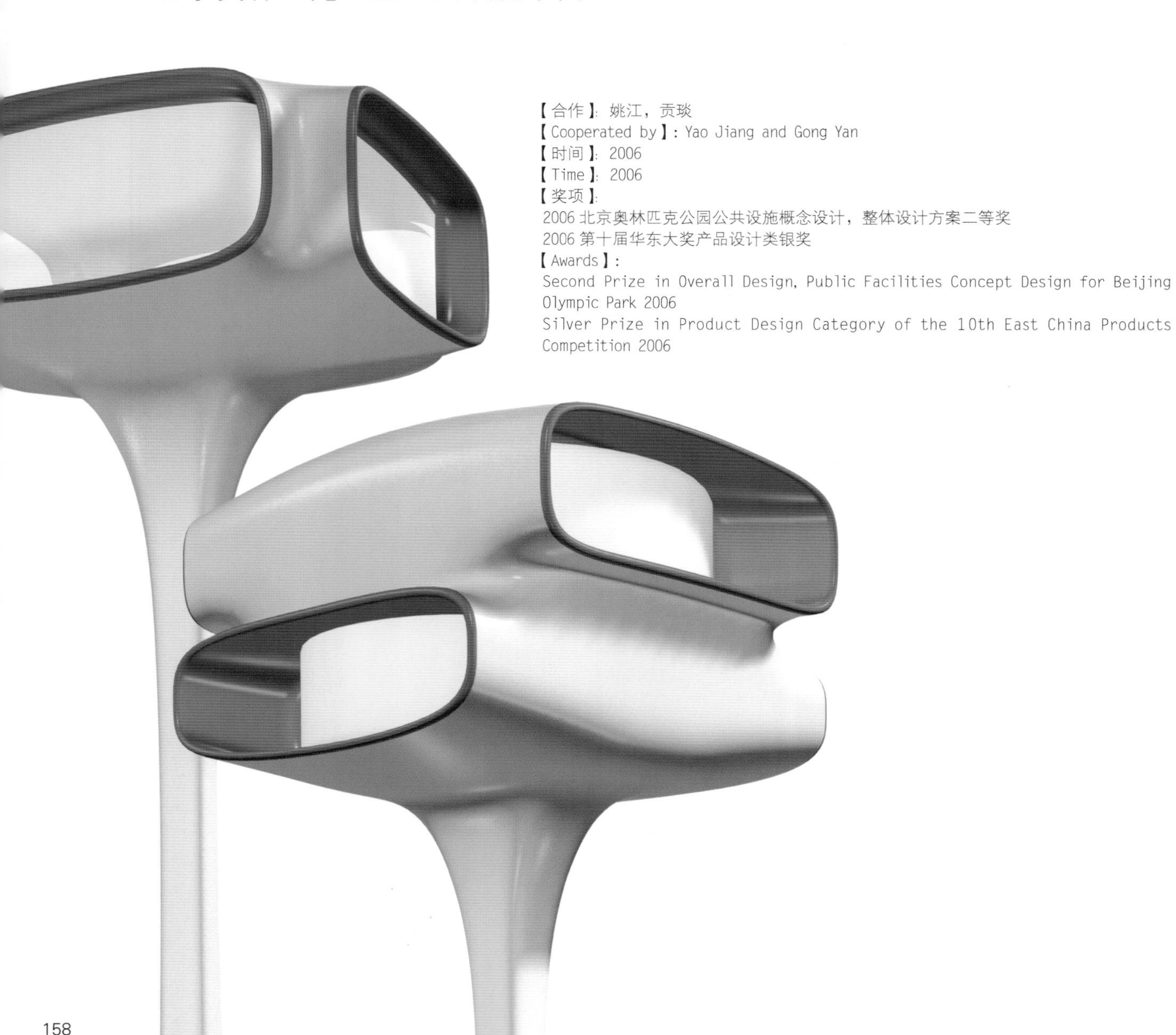

【合作】: 姚江，贡琰
【Cooperated by】: Yao Jiang and Gong Yan
【时间】: 2006
【Time】: 2006
【奖项】:
2006 北京奥林匹克公园公共设施概念设计，整体设计方案二等奖
2006 第十届华东大奖产品设计类银奖
【Awards】:
Second Prize in Overall Design, Public Facilities Concept Design for Beijing Olympic Park 2006
Silver Prize in Product Design Category of the 10th East China Products Competition 2006

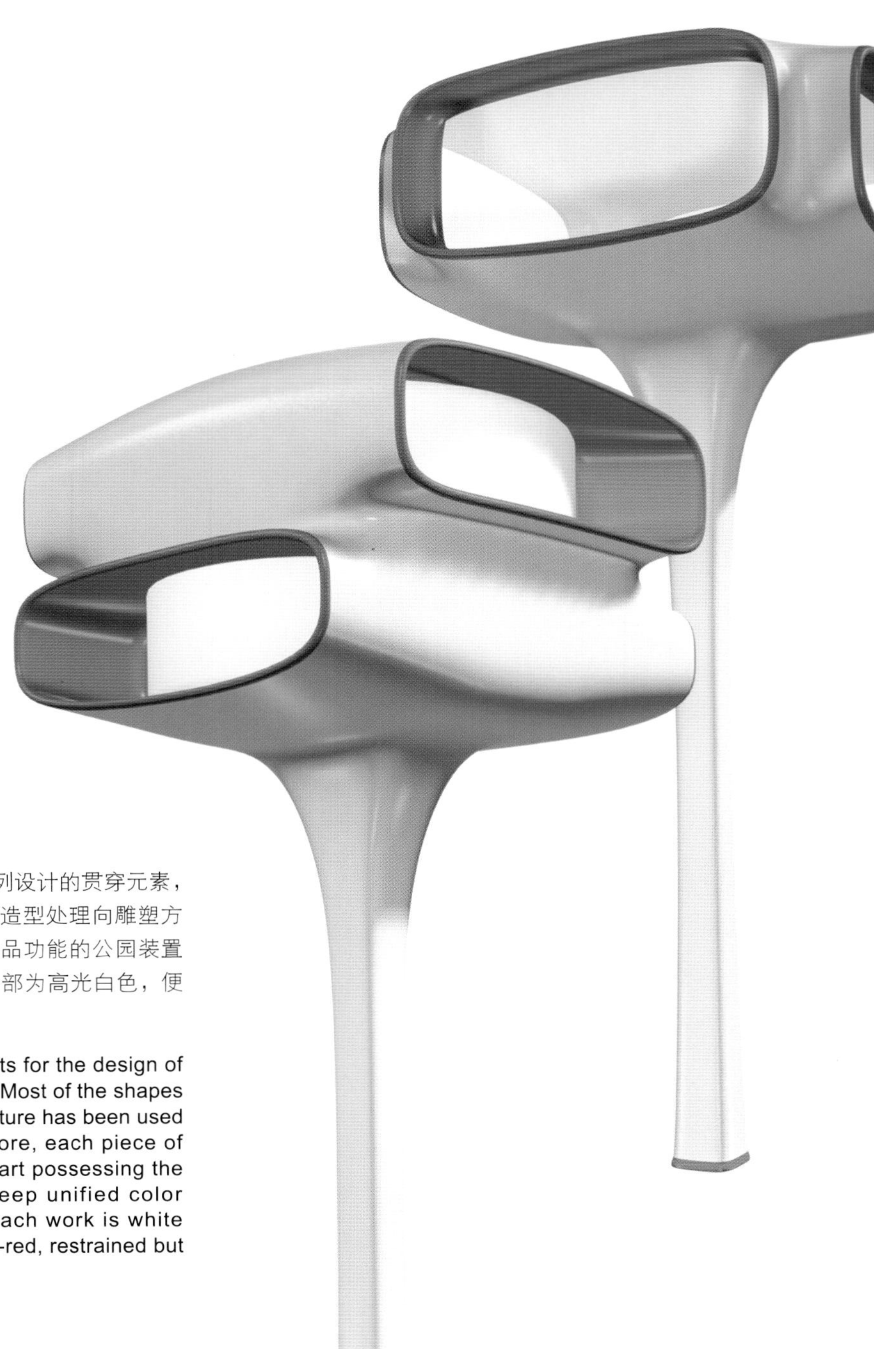

以有机形态作为北京奥林匹克公园公共设施系列设计的贯穿元素，形态多为圆润的曲面造型，并形成空腔体块，造型处理向雕塑方向有所借鉴，使得每件公共设施能成为具有产品功能的公园装置艺术。系列设计保持统一的色彩配置，形体外部为高光白色，便于清洗，内腔为艳丽的玫瑰红，内敛而奔放。

Organic forms have been taken as the elements for the design of public facilities series for Beijing Olympic Park. Most of the shapes are curves, and cavity blocks are formed. Sculpture has been used for reference in the shape treatment. Therefore, each piece of public facility can become a park installation art possessing the function of a product. The whole series keep unified color arrangement, and the external surface of each work is white polished, with its internal surface in bright rose-red, restrained but ardent.

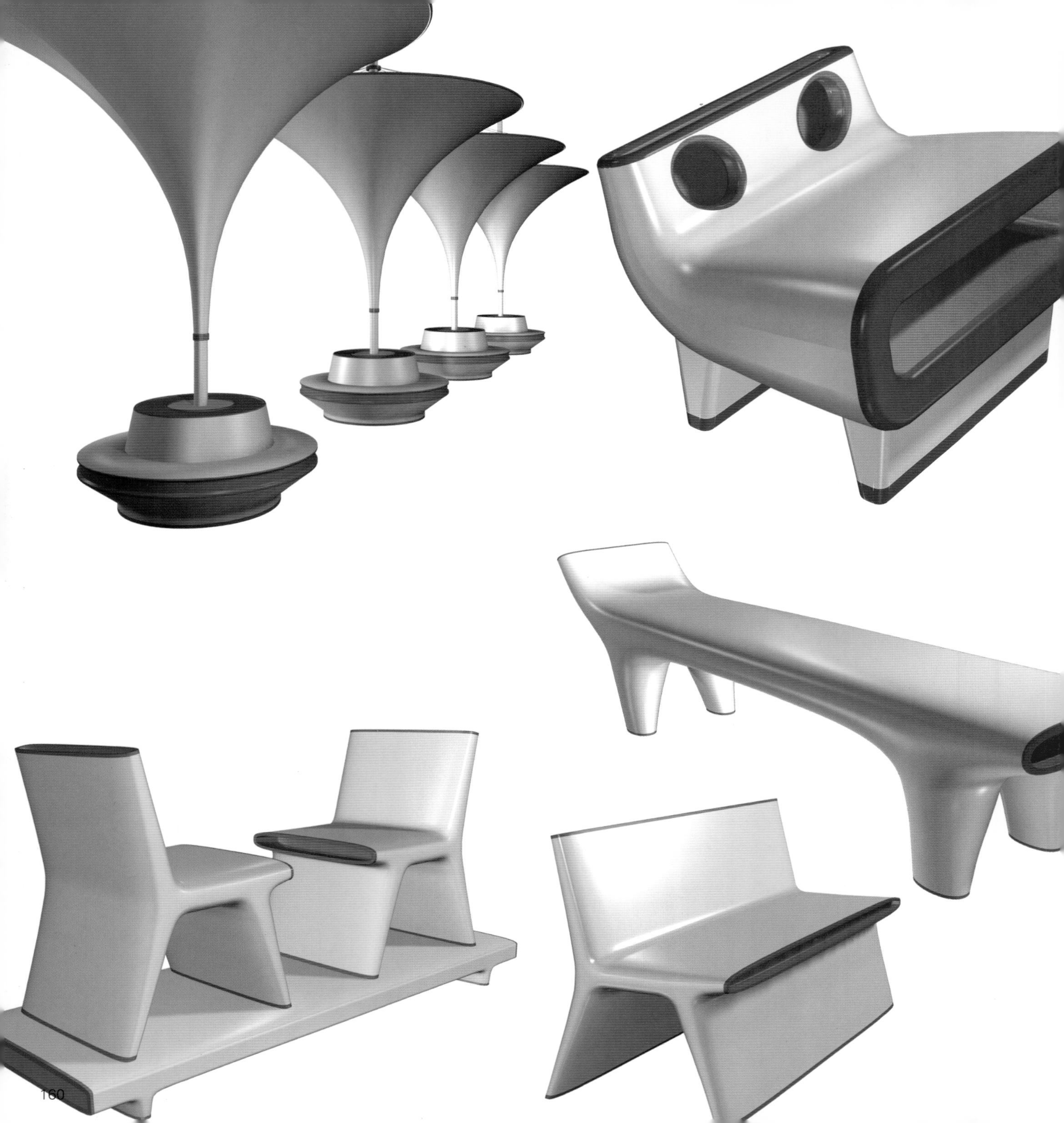

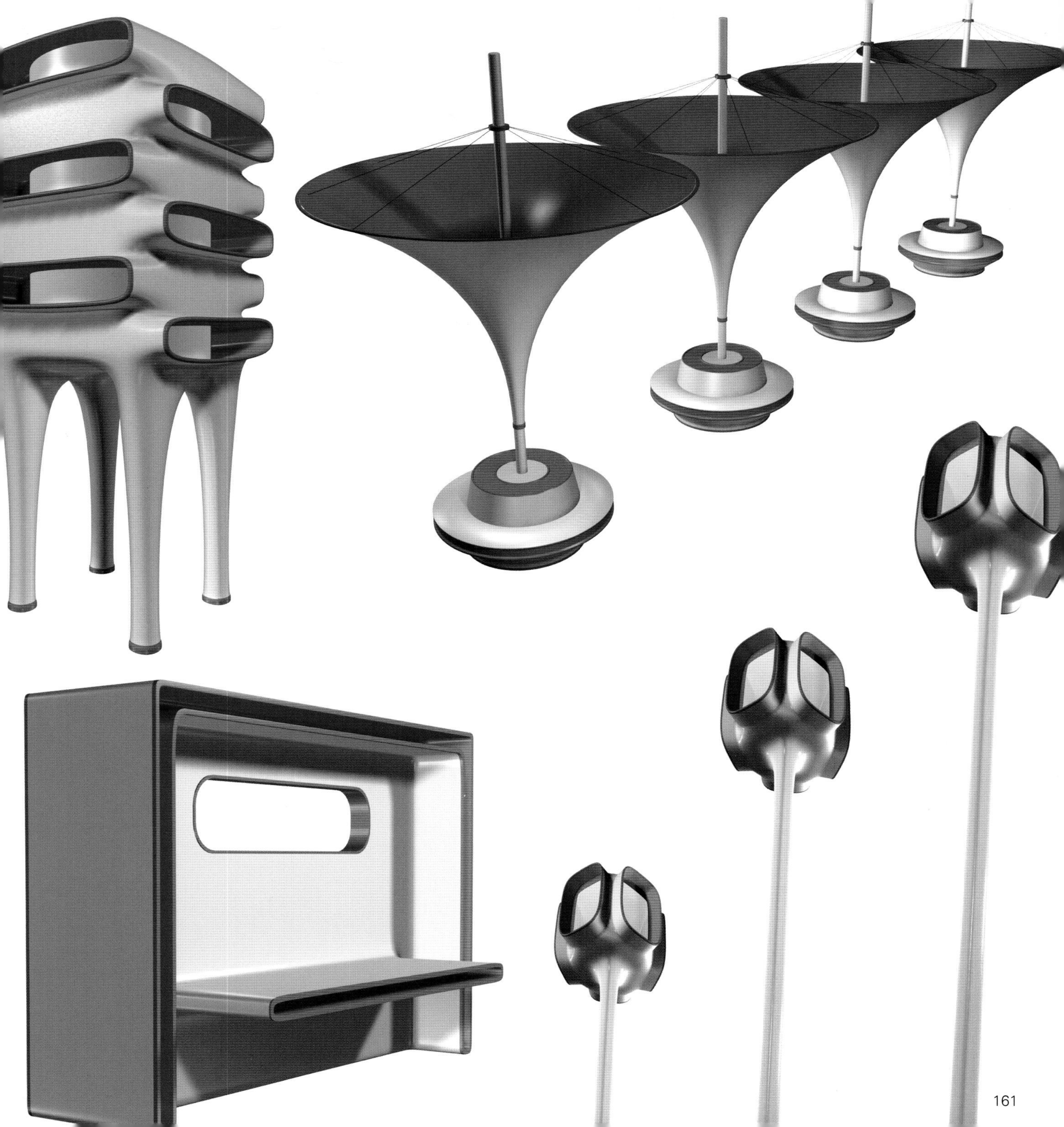

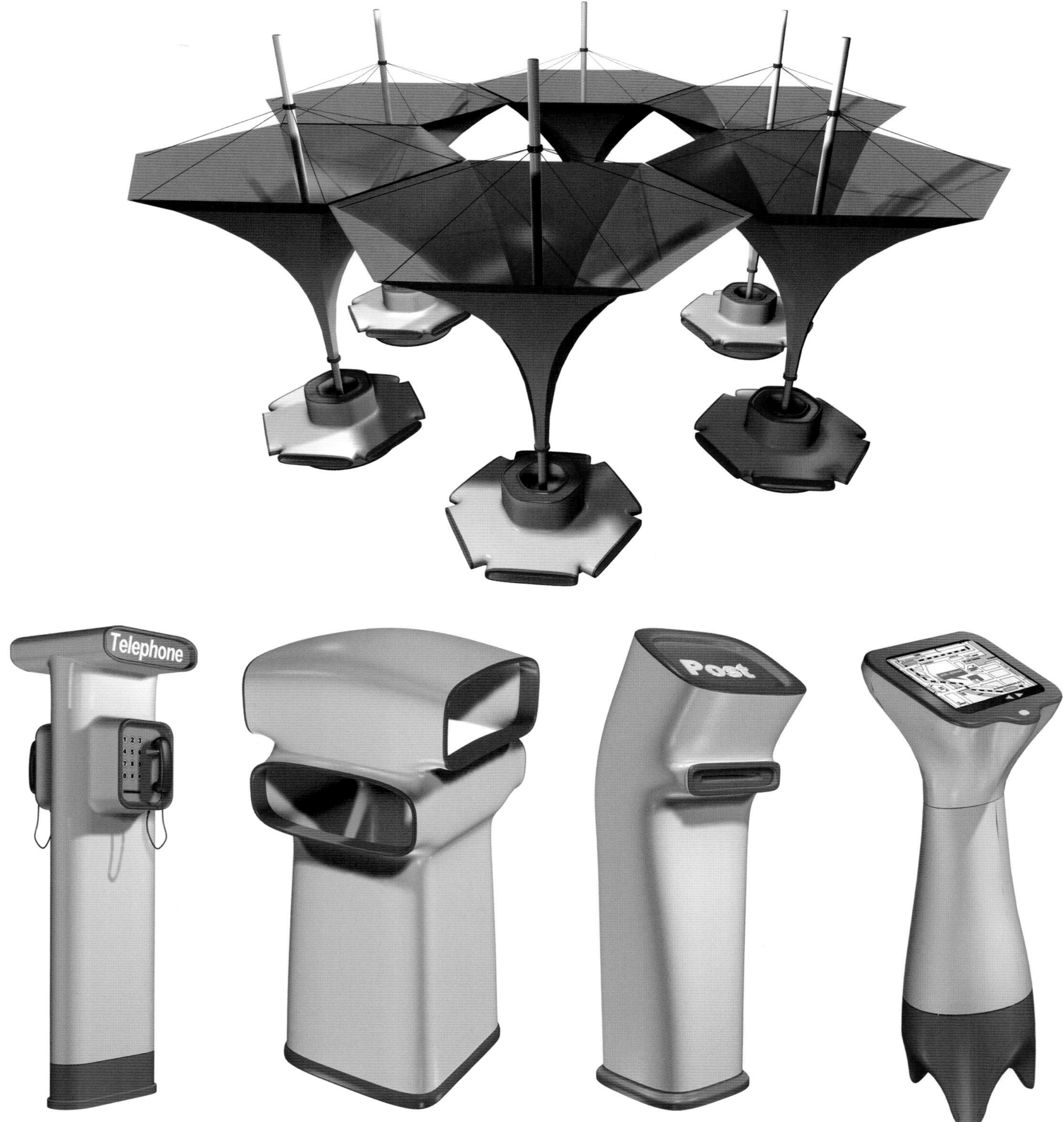
Telephone
Post

奥运会升旗广场 >>>
奥运会主会场 >>>
公园出口 >>>
<<< 地铁站入口
<<< 新安西大街
<<< 103 路公交站
Visit Map

照明与灯具
Lighting and Lamps

Flying Candle Holder
飞翔的烛台

【时间】：2007
【Time】：2007

为南京艺术学院 95 周年校庆设计的礼品方案，烛台两侧的翅膀使得烛台有展翅欲飞的的动感，材质为陶瓷。

This was a plan of gift designed for the 95th anniversary of Nanjing University of the Arts. The wings at both side of the holder makes it look like flying. The material is ceramic.

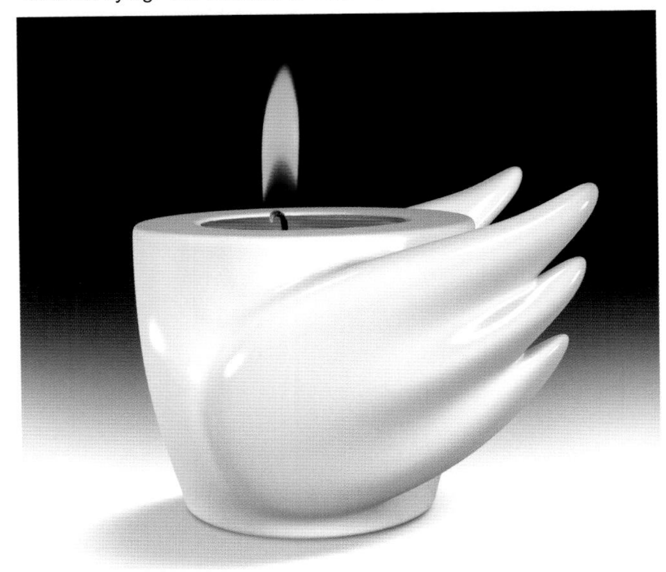

Candlestick on Mineral Water Bottle
矿泉水瓶上的烛台

【时间】: 2008
【Time】: 2008

此烛台可单独放置在桌面上，也可固定在普通矿泉水瓶上。烛台底部为突起部分，内有凹陷，配有与矿泉水瓶口径大小一致的螺纹，便于固定在矿泉水瓶上，以此提高烛台光亮的照射高度，同时便于手持。

This cadlestick can be not only placed on the desktop but also fixed in the ordinary mineral water bottle. In otder to be fixed on the mineral water bottle, the bottom of the candlestick is protruding with a thread that is fit with the mouth of the mineral water bottle, so that it can improve the irradition height of the candlestick light and make it easy to hold by hand.

Irregular Candle Holders
错落的烛台

【时间】: 2008
【Time】: 2008

三个不同高度的圆管成直角交织在一起形成一个烛台，圆管内径为蜡烛粗细，每两个圆管可以支撑为一个平稳的平面。可以根据圆管的不同长度确定烛台的高度。烛台为铝合金材质。

Three round tubes of different heights are put together to form a candle holder. The internal diameter is the size of a candle, and every two tubes can be used to support a stable plane. The height of the candle holder can be decided according to the different lengths of the tubes. The material is aluminum alloy.

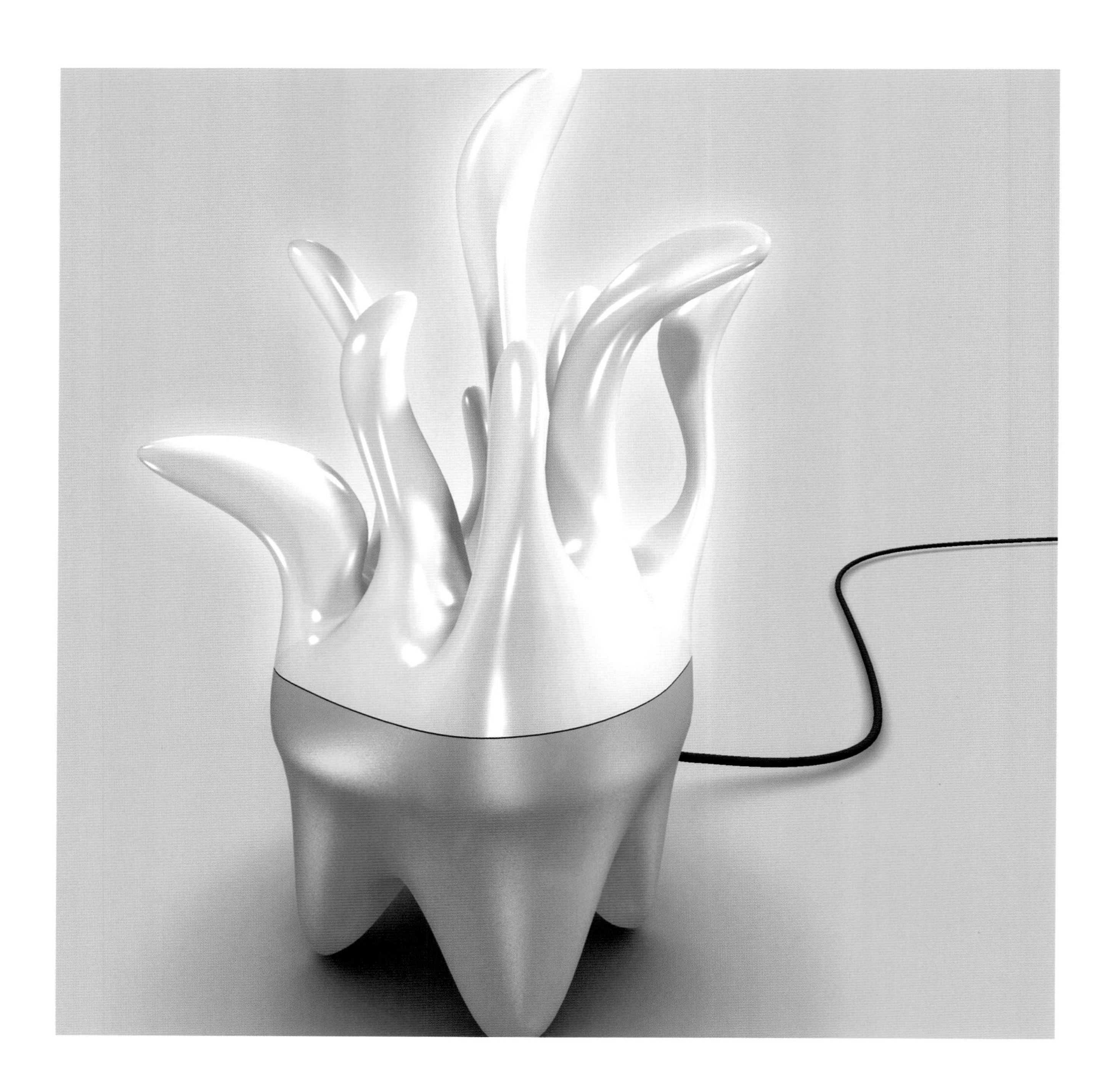

Lamp in Decayed Teeth Shape

蛀牙灯

【时间】：2007
【Time】：2007

去医院补蛀牙时想到的方案。牙齿造型的灯座中滋生出张牙舞爪的蛀虫，造型生动夸张。作品以鲜明的橙黄与白色相间，材质为 PVC 磨砂塑料，具有良好的透光性。

I got this idea when I went to hospital for my decayed tooth. The evil moth comes out from the lamp holder in tooth shape in a vivid and exaggerating pattern. Orange and white are used. The material is PVC matting plastic, with good light transmittance.

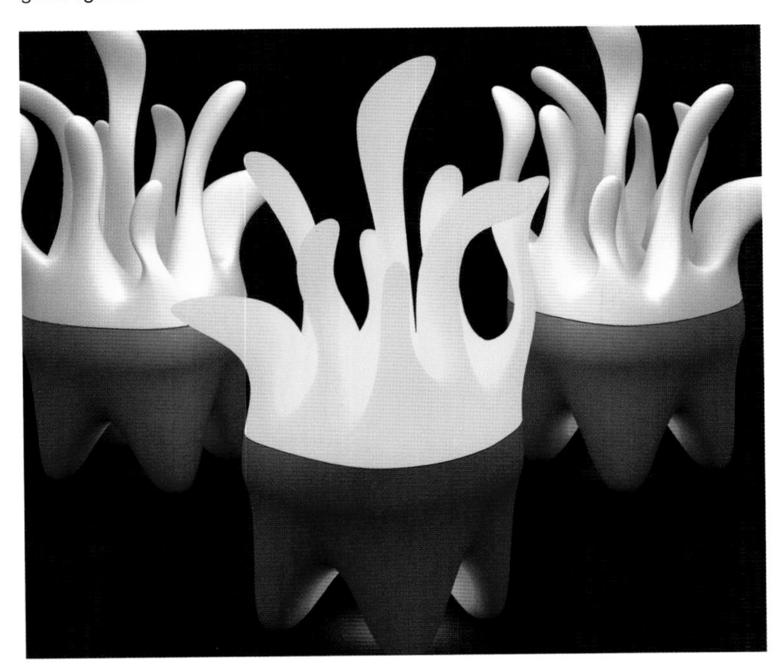

The Circle of Light
旋转的光

【时间】: 2007
【Time】: 2007

这是一个巧妙地将 LED 灯管与风扇融为一体的设计作品。吊扇的三个叶片为 LED 灯，当风扇开启的时候，光源随着叶片的旋转轻盈地舞动。电扇静止时，同样可以通过开关调节 LED 灯管亮度的强弱，成为一个吊灯。

This is one artful design work which combines LED and fan together in one article. Three blades on the ceiling fan are the LED light itself. As the fan turns on, the light source is running with the blades. When it stops, the LED turns into a normal hanging lamp with a switch to adjust the light brightness.

创新工具
Creative Tools

A Snail on the Map—A Map-measuring Band Tape
地图上的蜗牛——可测量地图的卷尺

【时间】: 2011
【Time】: 2011

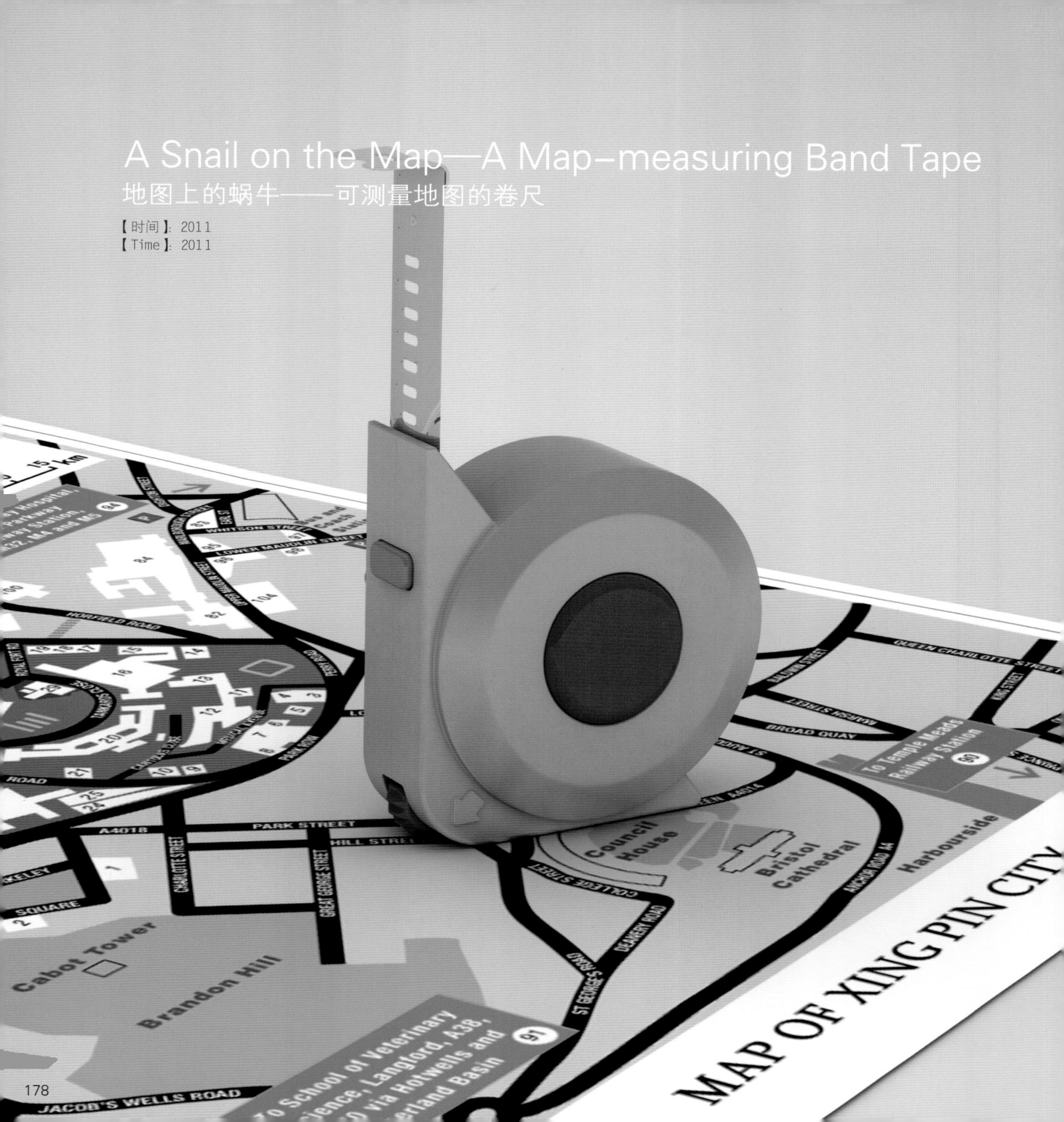

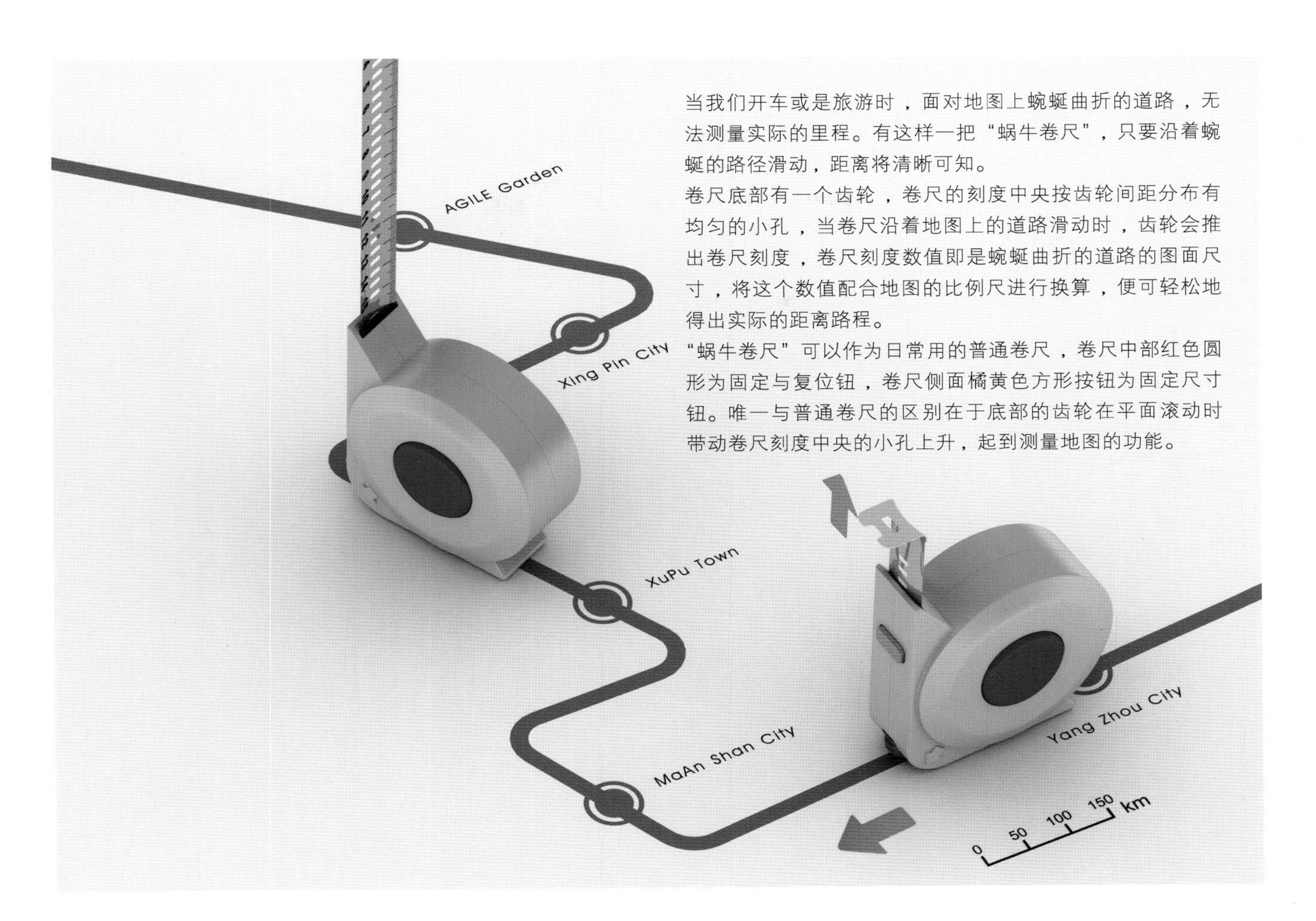

当我们开车或是旅游时，面对地图上蜿蜒曲折的道路，无法测量实际的里程。有这样一把“蜗牛卷尺”，只要沿着蜿蜒的路径滑动，距离将清晰可知。

卷尺底部有一个齿轮，卷尺的刻度中央按齿轮间距分布有均匀的小孔，当卷尺沿着地图上的道路滑动时，齿轮会推出卷尺刻度，卷尺刻度数值即是蜿蜒曲折的道路的图面尺寸，将这个数值配合地图的比例尺进行换算，便可轻松地得出实际的距离路程。

“蜗牛卷尺”可以作为日常用的普通卷尺，卷尺中部红色圆形为固定与复位钮，卷尺侧面橘黄色方形按钮为固定尺寸钮。唯一与普通卷尺的区别在于底部的齿轮在平面滚动时带动卷尺刻度中央的小孔上升，起到测量地图的功能。

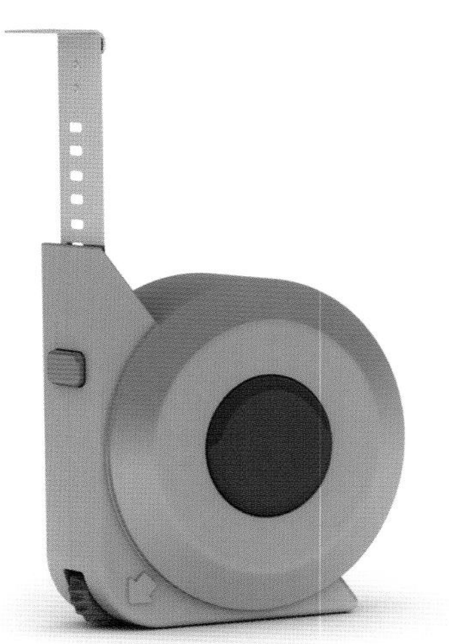

Facing winding roads on the map, we are unable to measure the actual mileage when we plan to drive cars or go on a trip. However, if you have such a "snail band tape", you can know clearly the distance by sliding the band tape along the winding path.

There is a gear wheel at the bottom of the "snail band tape" and there are small holes, in the cenerline of the graduation of the band tape, which are distributed at balanced intervals according to the wheel gauge. When the band tape is slided along the roads on the map, the wheel will push out the graduation of the band tape. The value of the graduation is the measure of the winding roads on the map. It is easy to extract the actual distance or mileage when the value is converted in conjunction with the scale of the map.

The "snail band tape" can serve as an ordinary daily-use band tape. The red round button in the center of the band tape item is a fixing and resetting button and the orange square button on the side of the band tape is a measure fixing button. Its only difference from an ordinary band tape is that the gear wheel at the bottom will push the holes in the centerline of the graduation of the band tape to rise when it rolls horizontally, serving as the function of measuring the map.

Grafting—Electrical Wire with the Cable Sockets

嫁接——配插座的电线

【时间】: 2006
【Time】: 2006

【奖项】: 入选 2008 中国包装技术协会设计委员会《中国设计年鉴》第六卷
【Awards】: Selected by Volume 6 of China Design Yearbook 2008, Design Committee of China Packaging Federation

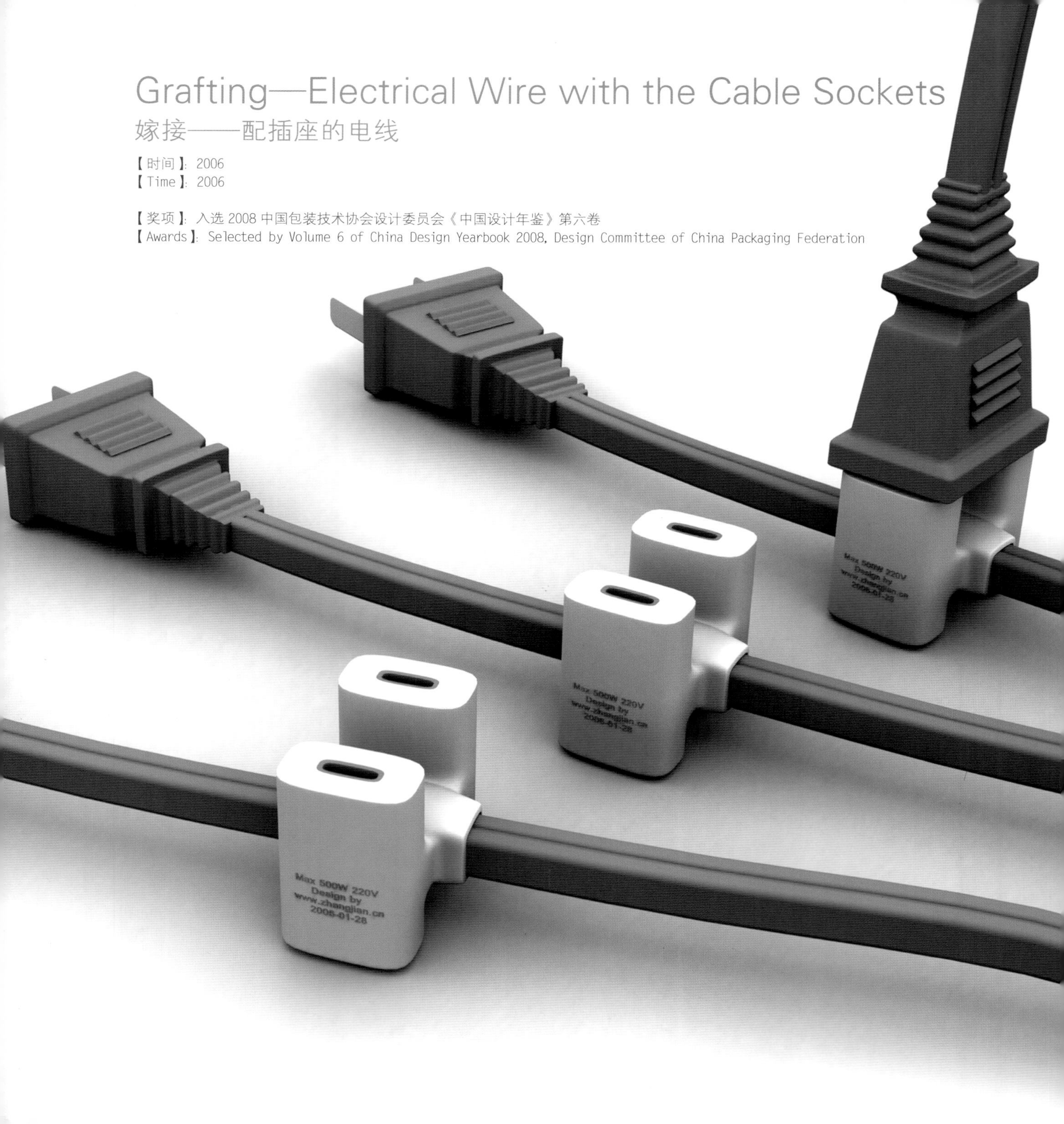

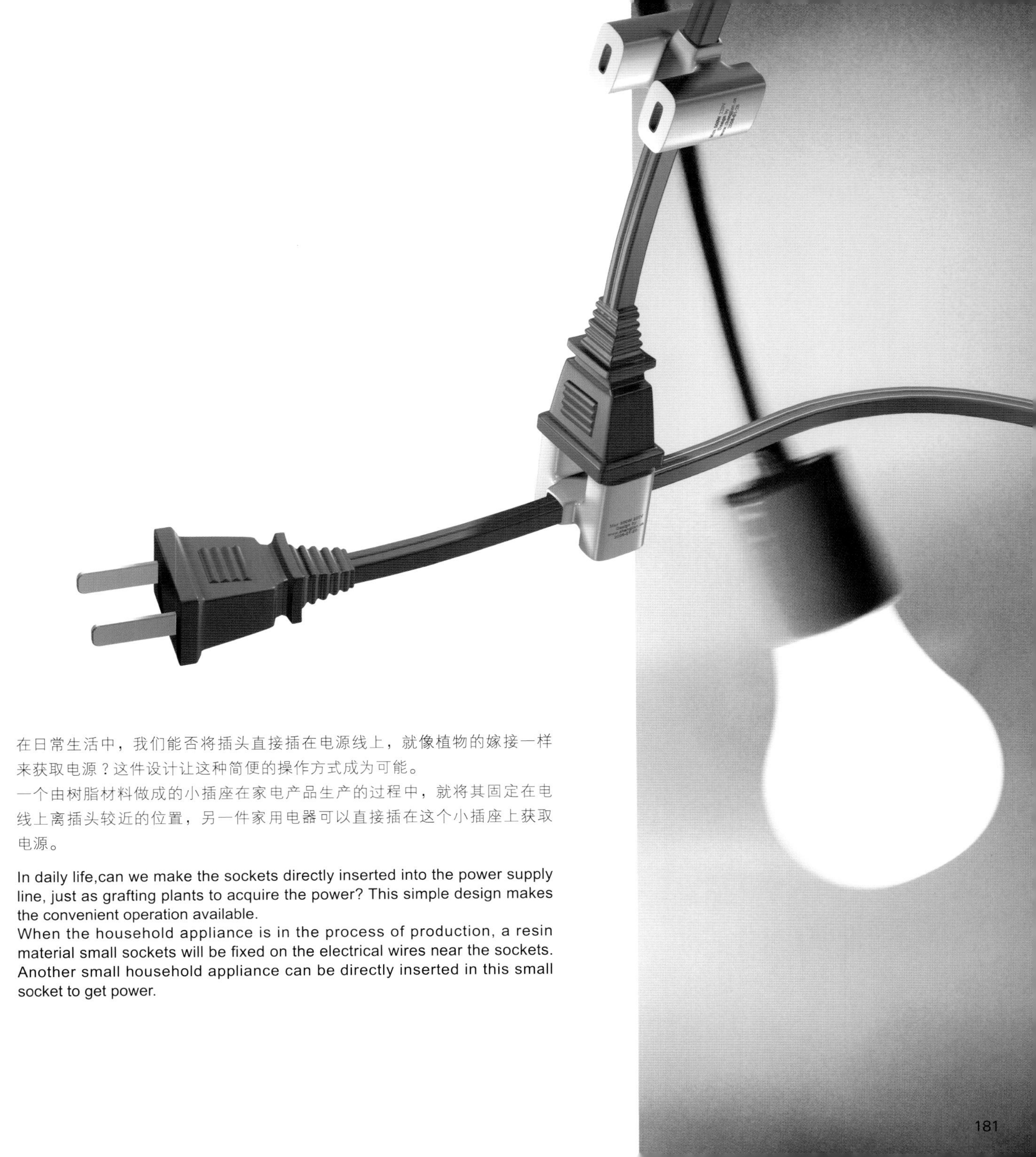

在日常生活中，我们能否将插头直接插在电源线上，就像植物的嫁接一样来获取电源？这件设计让这种简便的操作方式成为可能。
一个由树脂材料做成的小插座在家电产品生产的过程中，就将其固定在电线上离插头较近的位置，另一件家用电器可以直接插在这个小插座上获取电源。

In daily life,can we make the sockets directly inserted into the power supply line, just as grafting plants to acquire the power? This simple design makes the convenient operation available.
When the household appliance is in the process of production, a resin material small sockets will be fixed on the electrical wires near the sockets. Another small household appliance can be directly inserted in this small socket to get power.

Easy-hanging Cable
便于悬挂的电线

【时间】：2011
【Time】：2011

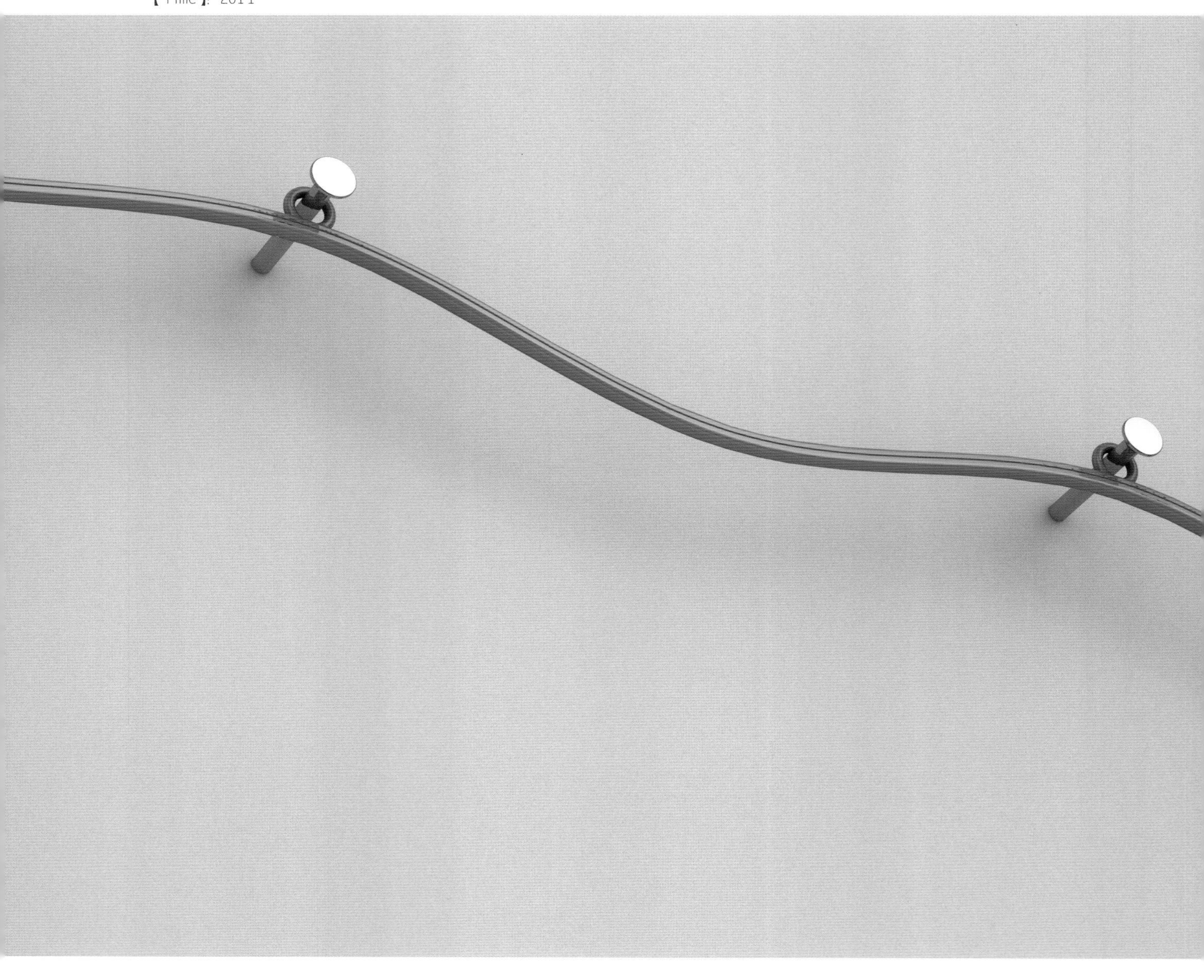

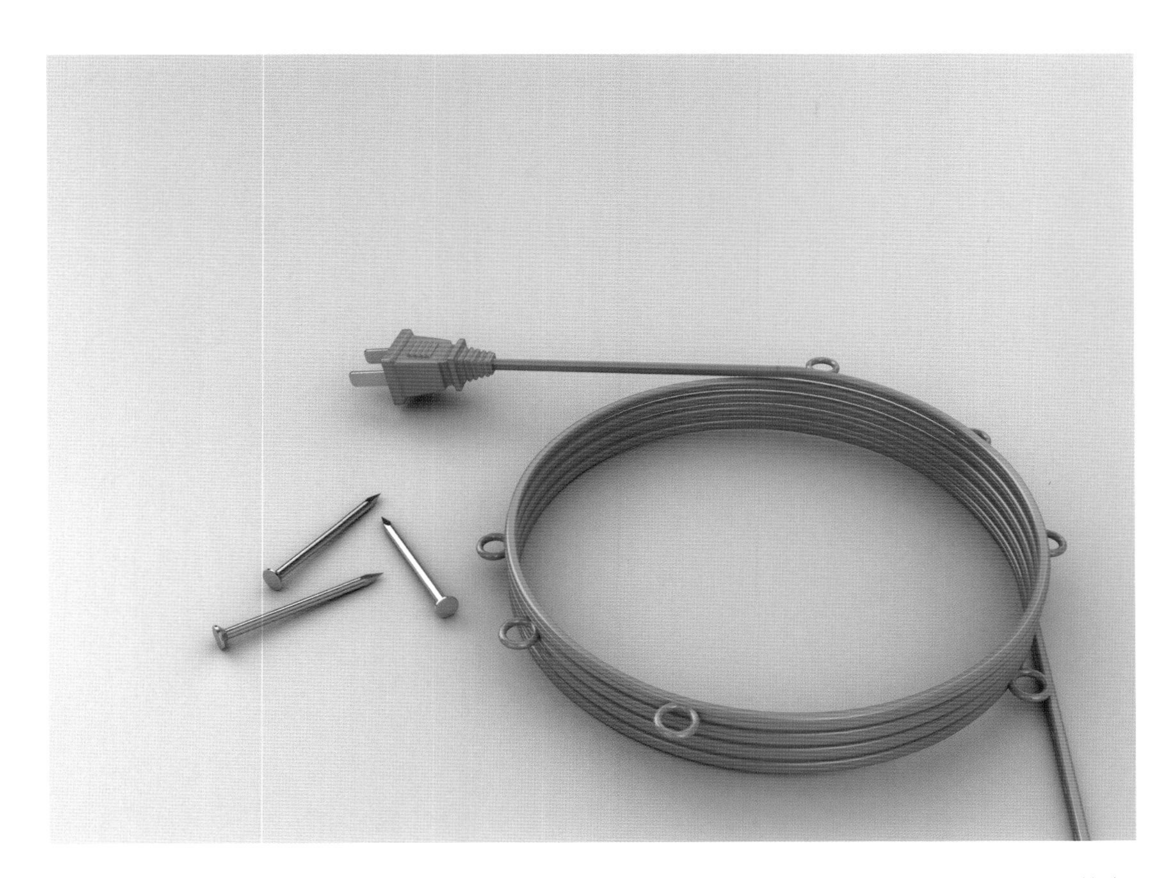

在一些施工场所中，电线遍布于地面，而我们解决这一问题多是将电线用胶布粘在墙面，不但不牢固而且繁琐。对电线作改良设计，生产电线时在电线每 50cm 的长度上附加一个圆环，钉子穿过这些圆环钉在墙面即可将电线悬挂起来，可轻松解决上述问题。

In some construction sites, cables are put all over the ground. How we usually solve this problem is to stick the cable onto the wall with tape, which is not stable and quite complicated. This design has improved the cable by adding a ring on the cable every 50cm, so that the cable can be hung on the wall with some nails to solve the problem mentioned above.

The Aperture Switch 光圈开关

【时间】：2009
【Time】：2009

【奖项】：
2009 韩国仁川国际设计比赛入选奖
入选 2010 中国包装技术协会设计委员会《中国设计年鉴》第七卷
【Awards】：
Selected Work of Incheon International Design Award 2009
Selected by Volume 7 of China Design Yearbook 2010, Design Committee of China Packaging Federation

利用相机光圈大小的变化特征，与家用可调节光源大小的开关结合，滑动开关环形拨盘可以调节开关中部光圈大小，灯的亮度与光圈的大小相配合，直观而有趣味。光圈内侧装有橘红色 LED 灯，便于在黑夜指示与使用。

The design combines the trait of the camera aperture size and the household adjustable light switch. Sliding the circular dial can adjust the aperture size in the middle of switch. The light brightness is relevant to the aperture size, which is intuitive and interesting. The inside of aperture is an orange LED lamp, for instructions and use at night.

Revolving Spiral

旋转的螺纹

【时间】: 2009
【Time】: 2009

此设计利用海螺纹样由中心向外延展，由小到大地旋转变量。旋钮安装在螺纹的轨道上。螺纹的中心为关闭，越向螺旋外延滑动旋钮，开关调节的亮度越大。

This design has utilized a conch spiral to extend from the center. The button is installed on the rail of the spiral, the center of which is off. The more you slide the button to the outside, the brighter the luminance will be.

Telephone Dialing Switch

电话拨盘开关

【时间】: 2009
【Time】: 2009

老式电话的拨盘使用方式与家用光源可调节开关结合，环形拨盘按照九个数字分为九个不同光源亮度。选择数字，将数字拨盘拨到卡口处数字选择的数字变亮，使用操作方式与老式电话一样。这件作品是对记忆中的老式电话的怀念，同时使用特点与老式电话有着很好的契合。

This is a combination of the old-style telephone dialing pad with the adjustable lighting switch for household use. The nine numbers on the pad indicates 9 levels of luminance. Move the pad to the clip, the selected number will shine. The way of operation is the same as the old-style telephone, the work shows the memory of which.

5
6
7
4
8
3
9
2
1
0
OFF

卫浴里的乐趣

Joy of bathing

11 Pieces of Series Designs for Add-ons of EAGO Sanitary Ware

益高卫浴附加产品系列设计 11 件

【合作】: 黄军花、刘芬、梁子宁、于庆庆
【Cooperated by】: Huang Junhua,Liu Fen,Liang Zining and Yu Qingqing
【时间】: 2010
【Time】: 2010
【奖项】: 2010 年广东省长杯工业设计大赛优秀奖
【Award】: Excellent Prize in Guangdong Governor Cup Industrial Design Competition 2010

此系列设计为童慧明教授与广东佛山益高卫浴企业签署的学院科研课题，以益高卫浴现有产品为基础，开发适合益高企业各类产品的附加品设计，使陶瓷卫浴产品更具生活气息，希望最终创建具有生活品质感的卫浴文化。与童慧明教授的三位研究生及南京艺术学院毕业生于庆庆合作完成。

The series of designs is a school research topic signed by Professor Tong Huiming and Guangdong Foshan EAGO Sanitary Ware Co., Ltd. The current products of EAGO have been considered as a basis to develop all kinds of add-ons suitable for the company to make the ceramic sanitary ware have more fun. The purpose is to create a sanitary ware culture with a sense of quality. I have completed the works with three postgraduate students of Professor Tong Huiming and Yu Qingqing, a student graduated from Nanjing University of the Arts.

“Happy Trojan”—Children Stool

“快乐木马”——儿童座便器

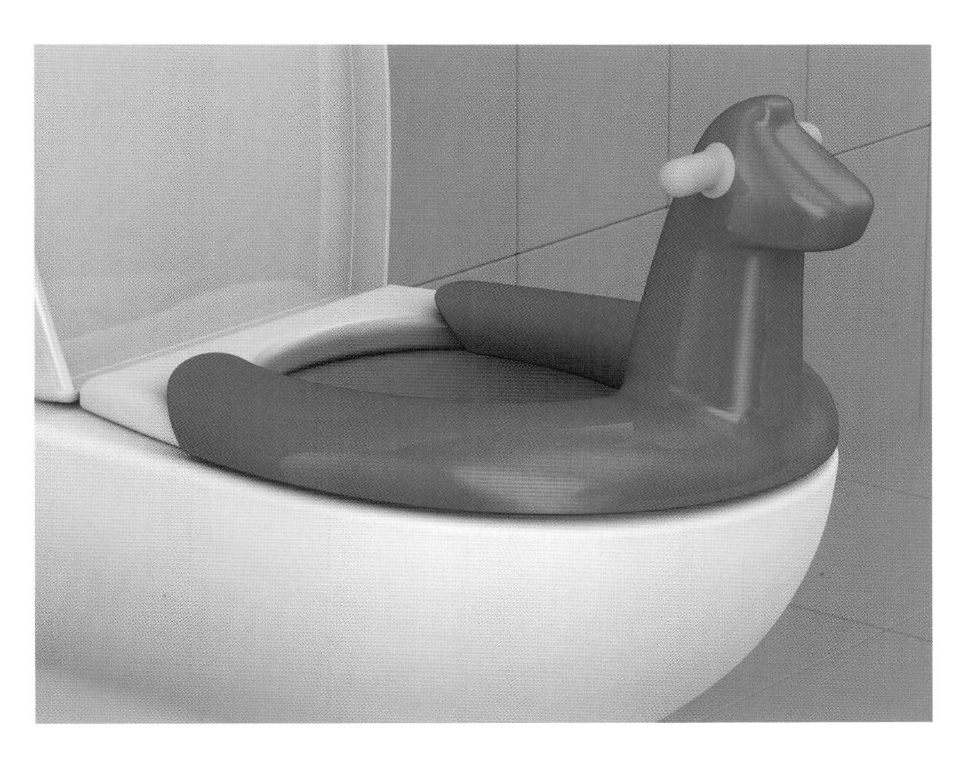

由于坐便器普遍为成人尺度设计，儿童不便使用，因此应用儿童木马的设计理念，从儿童的心理与身体尺度出发，将座便器设计成木马的造型，让儿童感到上厕所时像骑着游乐场里的木马，同时符合儿童身体尺度标准，便于儿童使用。造型简洁圆润，符合儿童用品的功能和心理特征。

The toilets are usually designed according to adults' sizes, not convenient for children' s use. Therefore, this design uses the idea of Trojan, based on psychological and physical standards of children to make them happy when they go to toilet. The shape is simple and round.

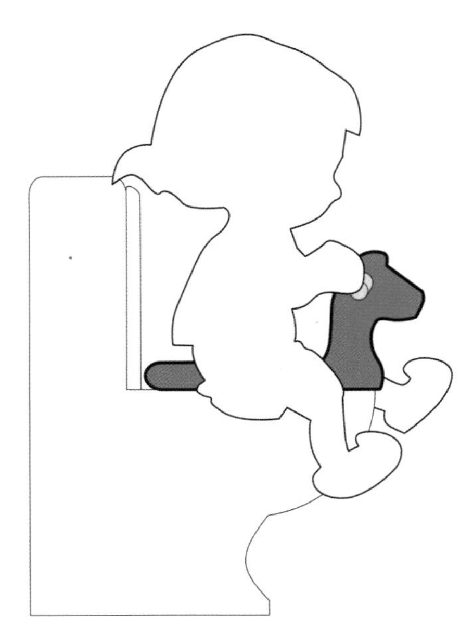

“POPO”—Toilet for Pets

“PO PO”——宠物座便器

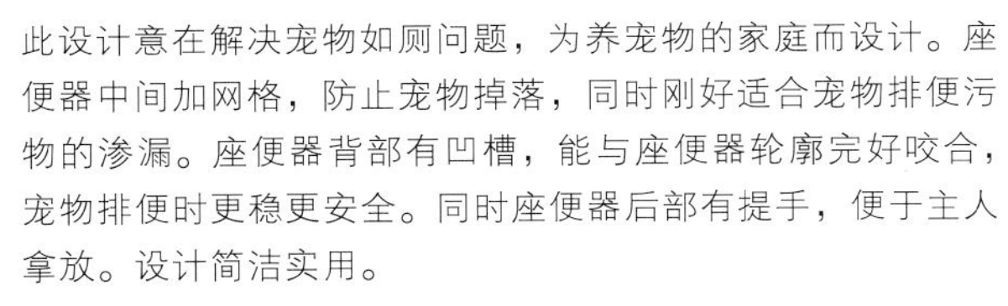

此设计意在解决宠物如厕问题，为养宠物的家庭而设计。座便器中间加网格，防止宠物掉落，同时刚好适合宠物排便污物的渗漏。座便器背部有凹槽，能与座便器轮廓完好咬合，宠物排便时更稳更安全。同时座便器后部有提手，便于主人拿放。设计简洁实用。

The purpose of this design is for the families that have pets. A grid is added in the toilet to avoid any pet dropping, meanwhile it is suitable for the seepage of pets' dejecta. On the back of this stool there is a groove to occlude with the toilet, making it stabler and safer for the pets. At the same time, there is a handle at the rear part of the stool for the pet' s owner to take. The design is simple and useful.

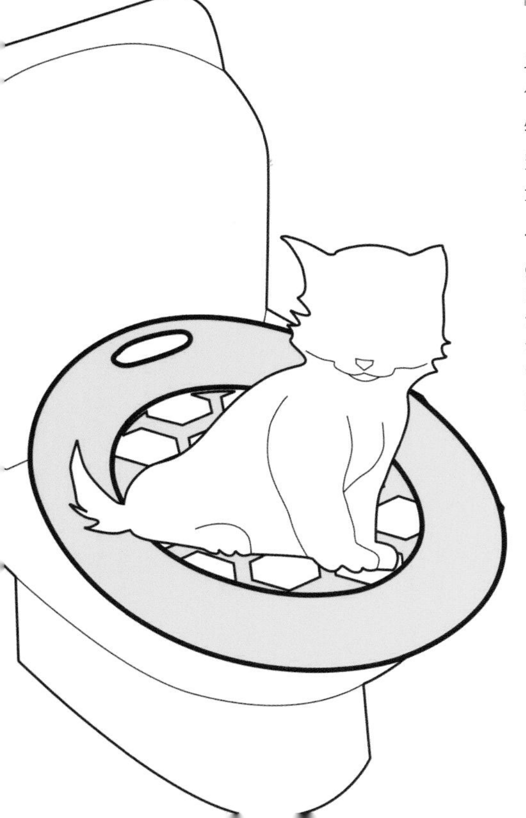

“Clothes”—Decorative Cover of Water Tank

“裳”——水箱装饰套

依据水箱的箱体上造型，为水箱制作一件“外衣”。纯净的白色上加上镂空的花纹，左侧有一体设计的花瓶造型，可用来插花，花瓶底部有旋钮可以换水。同时右侧可放置卫生纸筒。此附加件将功能与美观结合起来，将普通的马桶当做一件生活中的艺术品，具有简单而优雅的美感，营造出一种动人的生活情趣。

A “coat” is made for the water tank, which is pure white with pierced flowers patterns. On the left there is an integrated vase for putting flowers. On the bottom of the vase there is a knob for changing water. Toilet paper can be put on the right side. This product combines the function and beauty, and makes the normal toilet an artwork.

“Magic Bag”—Magazine Bag for Water Tank

“魔法袋”——水箱杂志袋

很多人上厕所有看杂志的习惯，此设计即在马桶水箱上附加杂志袋，可放置多本杂志，方便拿取，这是对卫浴空间的巧妙利用。杂志袋为布料材质，中间留出马桶出水按钮的位置。同样，该设计从生活细节出发，在不需要改变原有状况的前提下，通过附加件的巧妙设计，解决人们生活中的实际问题。

Many people like to read magazines when they use toilet. A bag for magazines is added on the water tank in this design, which is a smart use of the space. The material is fabric, on which the position for flushing button has been reserved. At the same time, this design has focused on details in life, having solved a real problem through a smart idea withont changing anything.

"Happy Family"—Children's Bath Chair

"亲子乐"——儿童洗澡椅

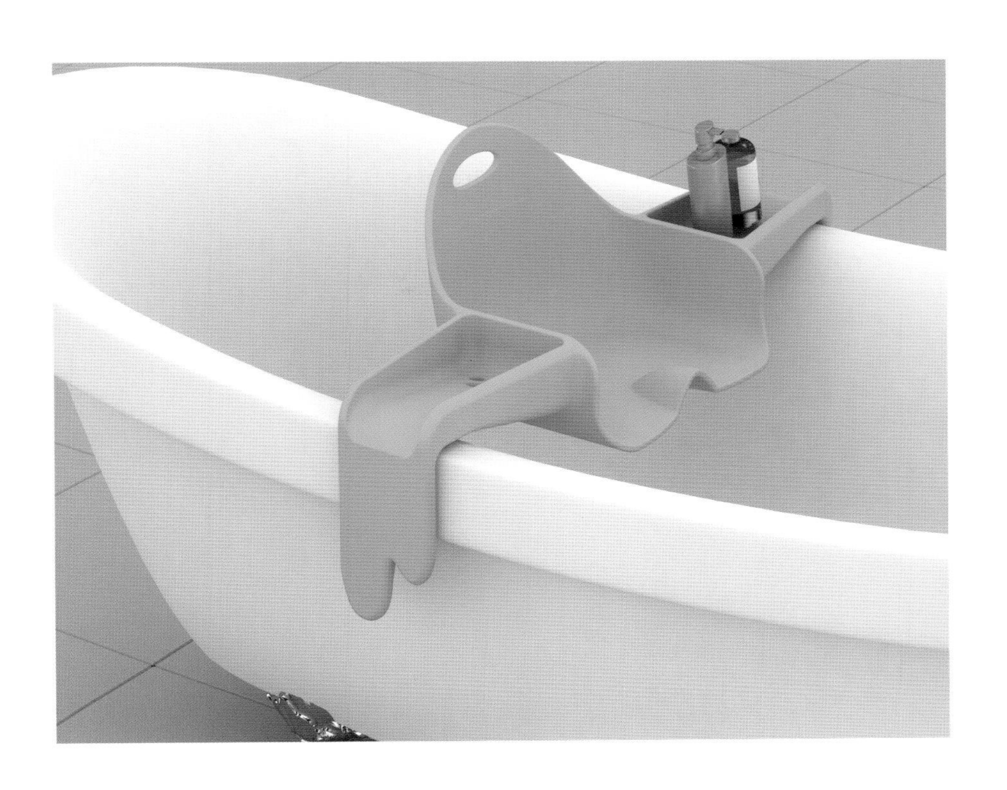

这件产品用于解决儿童洗澡问题。儿童洗澡时，将此座椅卡在浴缸边缘上，方便父母为儿童洗澡，椅子两侧有凹槽可以放置肥皂和洗浴液等物品，解决生活中的实际问题。

This product has solved the problem of taking a bath for children. It can be clamped on the edge of the bathtub so as to make it convenient for the parents to clean the kid. The slots on both sides of the chair can be used to put soap or shampoo, etc.

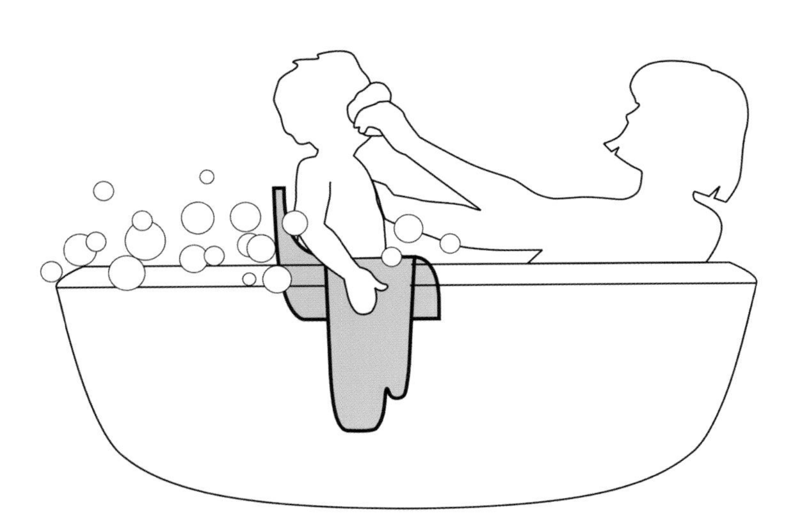

"Bear Rub"—Wall-hung Rubbing Pad

"熊搓"——挂墙式搓澡垫

由狗熊背靠大树搓痒联想到的设计——狗熊造型的大块搓澡垫，吸附在淋浴室墙上。搓澡垫由软橡胶材料制成，表面有许多突起颗粒肌理，可以涂上洗浴液供人在洗澡时搓背使用。用此款搓澡垫，可方便地清洁身体背部。整个设计将洗澡的动作变得憨厚有趣，同时又非常具有实用性。

The idea comes from a bear' s scratching on the tree. A rubbing pad in the shape of a bear has been designed to adhere to the wall of the shower. It is made of soft rubber, on which there are many protruding particles. People can apply shampoo onto it and rub the back of the body, convenient for cleaning the back. The design makes the shower more interesting, but useful as well.

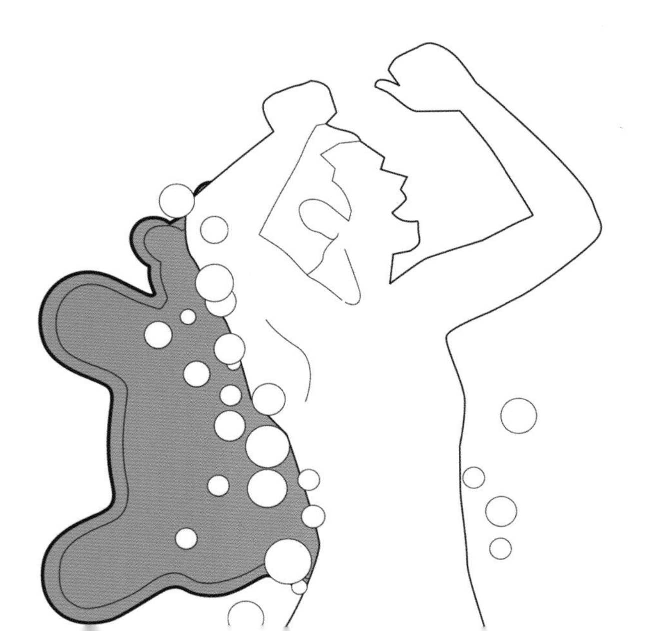

"Sunlight in Summer"—Reflecting Flowers

"夏日阳光"——反光花束

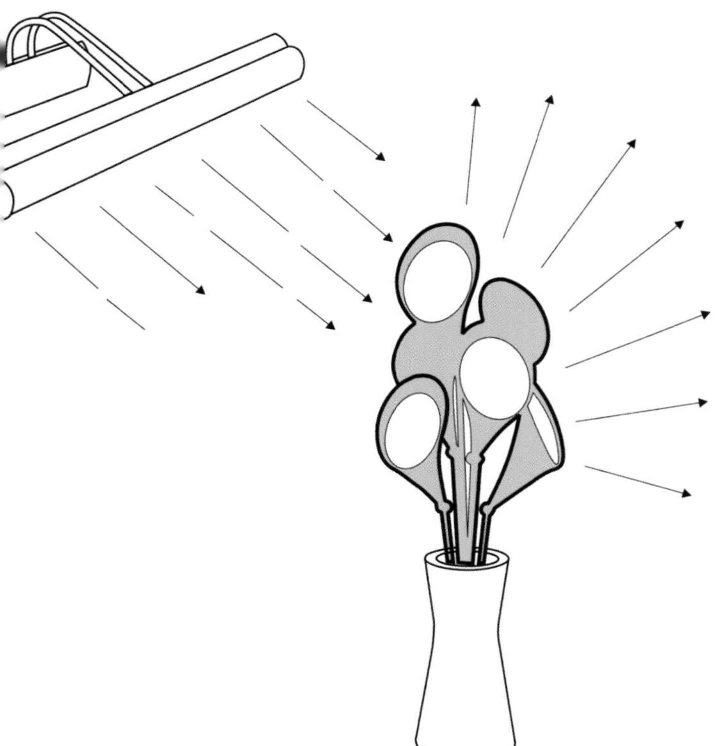

卫生间光线通常比较昏暗，利用凸面镜有反射并且放大光源亮度的特点，将一组凸面镜做成花束的造型，当光线照射过来时，通过调节各个凸面镜不同角度的反射，为浴室营造出光影斑斓的亮度和空间层次感，浴室的采光变得像夏日阳光照射进来一般绚丽。

Usually the bathroom is not very bright. By means of the features of convex mirrors, a set of which has been made into the shapes of branches of flowers. When light enters the bathroom, it will be reflected from different angles to form more brightness and spatial levels, making the bathroom as bright as in summer.

"Happy Family"—Bathroom Mirror
"全家福"——浴镜

这套"全家镜子"的设计，设计的核心是在普通的浴室配件——镜子中融入家庭和情感的设计理念。此套镜子为分别独立的镜面，镜面能方便地摘取，可根据家庭情况选择人数和家庭角色。同时，镜子外围加上画框，像一张浴室的全家福，家居环境更加温馨。此设计通过对普通镜子的改造，使得原本平淡无奇的浴室变得温情脉脉、和睦有趣。

The core of the design of this set of "mirrors for the whole family" is to add the concept of family and emotion into mirrors—the normal bathroom accessories. The set of mirrors comprises independent ones, which can be easily taken down. People can choose one according to the number and roles of family. At the same time, a frame is added to the mirror to make it looks like a photograph of the whole family, offering a cozy feeling.

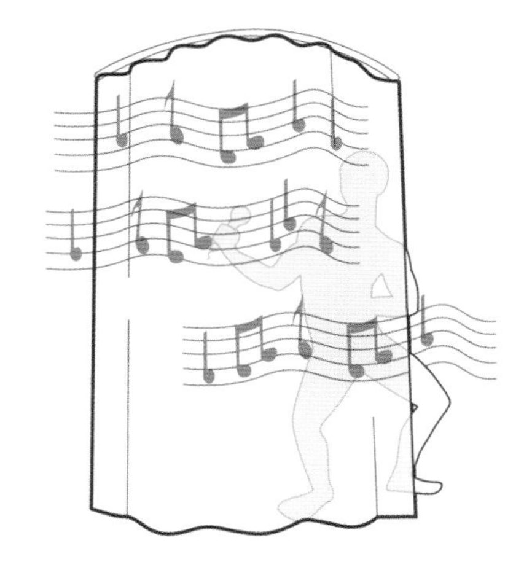

"Great Singer"—Music Shower Curtain
"麦霸"——音乐浴帘

在浴帘上印上歌曲的音符。可以一边洗澡，一边看着浴帘，哼自己爱的歌曲，轻松洗去一天的疲劳，生活小乐趣无处不在。同时，浴帘成本低廉，使用周期较短，可随心更换各种歌曲，并方便摘取。

Musical notes are printed on the shower curtain. People can sing the song they like when taking a shower. The cost is very low and the use period is relatively short. Songs can be changed as well.

“Kalier”—Basin Washboard

“卡丽尔”——台盆搓衣板

依据台盆的弧度造型，在搓衣板的背部设计凹槽，可用于搓衣板在台盆上的固定。此设计解决了实际的生活问题，为面盆增加了新的功能,同时并未增加过多成本。造型简单，构思巧妙，成本低廉。

According to the arc of the basin, a groove has been designed on the back of the washboard to fix it on the basin. This design has solved an actual problem in life and added more function to the basin without increasing much cost.

“Carlo”—Basin Pad

“卡洛”——台盆水漏

这款台盆水漏，意在解决用台盆清洗衣物时衣物的搁置问题。我们经常在台盆上洗衣物时，衣物无处可放。此附加件背后有凹槽设计，可卡在台盆上用于固定，中间有渗水的漏孔。此设计从细节出发，解决人们生活中的不便之处。简单、却巧妙实用。

This basin pad with slots for water is to solve the problem of putting clothes when we wash clothes. Sometimes we wash some clothes by hand in the basin but there is no place to put them. This pad can be clamped onto the basin with holes on it to drain water. It has focused on the details in life.

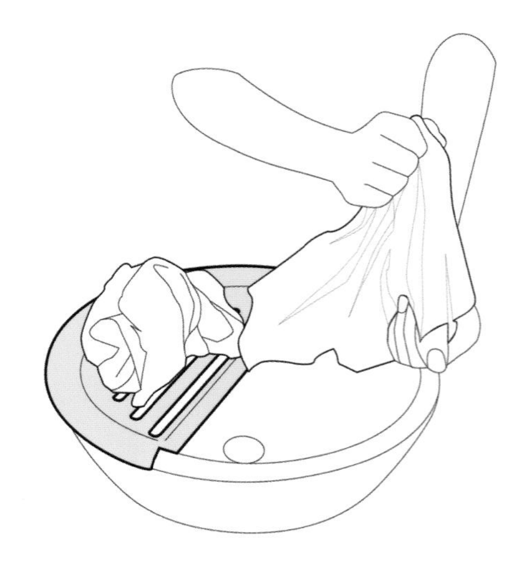

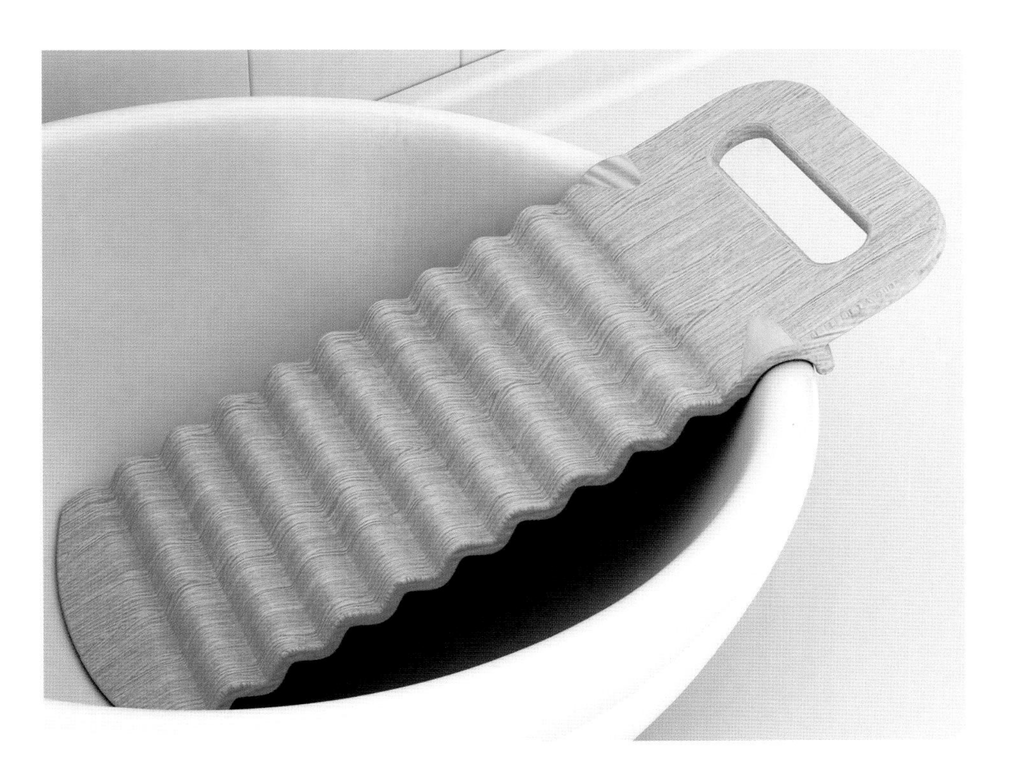

Mirror in the Clouds
高耸入云的镜子

【时间】: 2010
【Time】: 2010

极具装饰感的祥云纹样镜子，悬挂在卫生间的瓷砖墙面上。地面有一个梯子与云形镜子连接，好似可以从地面攀爬到云端，形成一个趣味而浪漫的场景；当然，梯子的横档有一定的功能，可以挂衣物或毛巾。

The very decorative mirror in the cloud shape is hung on the ceramic wall of the bathroom. There is a ladder from the ground connecting to the mirror, as if people could climb into the clouds, which is interesting and romantic. The crosspieces of the ladder also have certain functions, such as hanging clothes or towels.

The Bible Mirror
圣经镜子

【时间】：2009
【Time】：2009

一本《圣经》，悬挂于墙上，当我们以虔诚之心开打时，《圣经》里看到的竟是我们自己。《圣经》书籍外壳为软质橙色橡胶，易于弯折。镜子外框为白色塑料。

When we open the Bible hung on the wall with a sincere heart, we see ourselves in the mirror. The case of the mirror is made of soft rubber in orange, easy for bending. The frame of the mirror is white plastic.

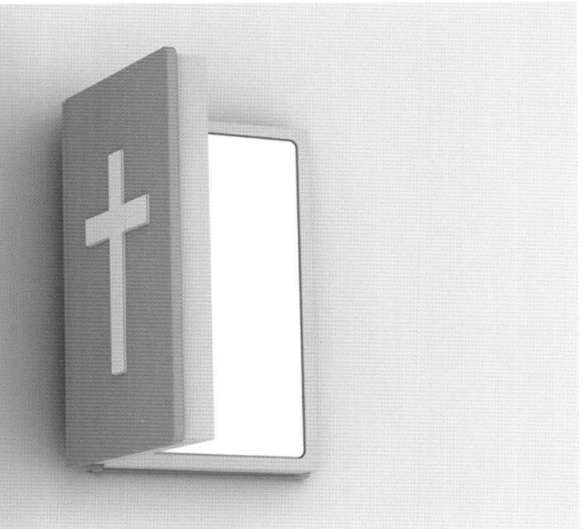
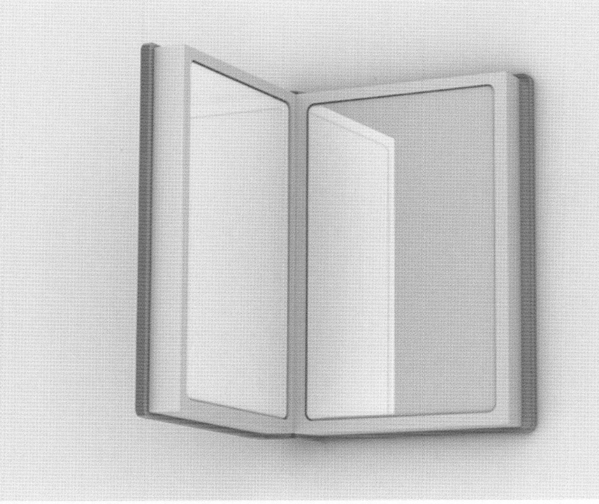

Out of Door
门外

【时间】: 2010
【Time】: 2010

在塑料门框造型内，利用开启的门缝的虚空间设置一面镜子，当我们照镜子时会带着一种期盼与幻觉意识，希望在开启的门缝中发现期盼的景物，这种期盼的趣味心理是这件作品设计的创意出发点。

Inside the plastic frame, a mirror is set by means of the imaginary space of the opening door. We might have some expectation or get some illusion in front of the mirror, hoping to find something out from the seam of the door. Expectation is the idea of this design.

Hanging Mirror in Reindeer Shape

驯鹿挂镜

【时间】: 2010
【Time】: 2010

很普通的塑料框架镜子，在框架的一端附加一只驯鹿的半个头颅，透过镜子则可以折射出完整的驯鹿头形象。而设置驯鹿头的目的不仅是装饰，驯鹿头延展出的华丽鹿角可以悬挂毛巾或衣物。

This is a very normal mirror in a plastic frame, only half of the head of a reindeer has been added on one side of the frame. By the reflection of the mirror, the image of the head of a reindeer will appear. The purpose of the reindeer' s head is not only for decoration, but also for hanging towels or clothes.

论文获奖

2012 年由中国工业设计协会指导，《设计》杂志社和中国工业设计协会专家工作委员会主办的“中国工业设计发展十年优秀学术论文评选”，张剑共 5 篇论文获奖：

《设计创新对不良产业链良性引导的可行性探析——以“黑心棉”为例》获银奖

《对“家具之上”的解读——由广美家具专业毕业设计看小众化设计与个性化培养》获优秀论文奖

《走下 T 台的低碳设计秀》获优秀论文奖

《由诗歌的意象手法看产品设计语义的发展》获论文奖

《传统元素在产品设计中的“去标签化”讨论》获论文奖

作品获奖

2010

《儿童储物椅》获 2010 德国国际红点设计概念奖

《卫浴附加产品系列设计》（合作）获 2010 年广东省长杯工业设计大赛优秀奖

《整合卫浴空间设计》（合作）获 2010 年广东省长杯工业设计大赛鼓励奖

《立体拼图家居用品系列》入选中国包装技术协会设计委员会《中国设计年鉴》第七卷

《Here My home》入选中国包装技术协会设计委员会《中国设计年鉴》第七卷

《红鲤与鳞片——零钱储蓄》入选中国包装技术协会设计委员会《中国设计年鉴》第七卷

《儿童储物椅》入选中国包装技术协会设计委员会《中国设计年鉴》第七卷

《儿童组合椅凳》入选中国包装技术协会设计委员会《中国设计年鉴》第七卷

《坠机磁贴》入选中国包装技术协会设计委员会《中国设计年鉴》第七卷

《树形钥匙挂钩》入选中国包装技术协会设计委员会《中国设计年鉴》第七卷

《牛头与皇冠酒瓶塞》入选中国包装技术协会设计委员会《中国设计年鉴》第七卷

《光圈开关》入选中国包装技术协会设计委员会《中国设计年鉴》第七卷

《江苏沃得装载机》入选中国包装技术协会设计委员会《中国设计年鉴》第七卷

《烟囱烟灰缸》入选中国包装技术协会设计委员会《中国设计年鉴》第七卷

《工作时间》（合作）入选中国包装技术协会设计委员会《中国设计年鉴》第七卷

《带刻度的桌子》（合作）入选中国包装技术协会设计委员会《中国设计年鉴》第七卷

2009

《年轮卷纸》（合作）获 2009 韩国仁川国际设计比赛银奖

《光圈开关》获 2009 韩国仁川国际设计比赛入选奖

《红鲤与鳞 – 零钱储藏》入选 2009 Sparkawards 星火国际设计奖

《环保花瓶》第十一届全国美展，获提名奖

《烟囱 – 烟灰缸系列》获“2009 创新顺德国际工业设计大赛”专业组铜奖

《儿童储物椅》获 2008 江苏优秀工业设计二等奖

《绿色的凳子》（合作）入选中央美术学院“第四届为坐而设计”比赛

《坠机 – 磁贴》入选意大利“Beyond Silver”国际设计比赛

《牛头和皇冠 – 瓶塞》入选意大利”Beyond Silver“国际设计比赛

2008

《工作空间》(合作)获 2008 韩国仁川国际设计比赛特选奖

《儿童储物椅》获 2008 韩国仁川国际设计比赛入选奖

《鹤鹿同春雕塑》(合作)入选中国包装技术协会设计委员会《中国设计年鉴》第六卷

《系列卡通造型》入选中国包装技术协会设计委员会《中国设计年鉴》第六卷

《我的背后—椅子》入选中国包装技术协会设计委员会《中国设计年鉴》第六卷

《嫁接—配插座的电线》入选中国包装技术协会设计委员会《中国设计年鉴》第六卷

《可以调节关系—长椅与茶几》入选中国包装技术协会设计委员会《中国设计年鉴》第六卷

《琉璃小人书立》入选中国包装技术协会设计委员会《中国设计年鉴》第六卷

《旗袍椅》入选中国包装技术协会设计委员会《中国设计年鉴》第六卷

《酌—酒器》入选中国包装技术协会设计委员会《中国设计年鉴》第六卷

《骑驴—椅子》入选中国包装技术协会设计委员会《中国设计年鉴》第六卷

《废弃自行车家居系列》入选中国包装技术协会设计委员会《中国设计年鉴》第六卷

2007

《洒水壶花瓶 》入选意大利 “ceramics for breakfast ” 国际设计比赛

《我的背后 – 椅子》获 2007 北京 798 “旋转的硬币” 国际工业设计展优秀作品奖

《空中芭蕾 – 喂鸟器》获 2007 北京 798 “旋转的硬币” 国际工业设计展优秀作品奖

《猫咪的玩具》获 2007 北京 798 “旋转的硬币” 国际工业设计展优秀作品奖

《够草帽的小人》获 2007 北京 798 “旋转的硬币” 国际工业设计展优秀作品奖

《印迹 – 2007 拖鞋》获 2007 北京 798 “旋转的硬币” 国际工业设计展优秀作品奖

《充气的吊灯》获 2007 北京 798 “旋转的硬币” 国际工业设计展优秀作品奖

《骑驴椅子》获 2007 北京 798 “旋转的硬币” 国际工业设计展优秀作品奖

《弹簧淋浴空间》获 2007 北京 798 “旋转的硬币” 国际工业设计展优秀作品奖

《可拖动的悬空支架》获 2007 北京 798 “旋转的硬币” 国际工业设计展优秀作品奖

2006

《北京奥林匹克公园公共设施系列部分》(合作)获第十届华东大奖产品设计类，银奖

《果实 – 鸟窝设计》获第十届华东大奖产品设计类，优秀奖

《废弃自行车家具系列产品》(集体)获 2006 中国家具设计大赛三等奖

《北京奥林匹克公园公共设施概念设计》(集体)获北京奥林匹克公园整体设计方案二等奖

2005

《骑驴－椅子》获“嘉宝莉漆杯2005中国家具设计大赛”专业组优秀奖
《充气的吊灯》获“嘉宝莉漆杯2005中国家具设计大赛”专业组鼓励奖
《酌（酒瓶＋酒杯）》获“改变包装”国际设计连线全球竞赛最高奖——全场大奖
《报纸沙发》入选中央美术学院第二届“为坐而设计”比赛
《A Door Design for All Members of the Family》入选意大利“A Door to Paradise”国际设计比赛
《1+1+1》模块化庭院空间 获2004日本Designtop“庭院空间国际设计比赛”优秀奖
《弹簧淋浴空间》入选中国包装技术协会设计委员会《中国设计年鉴》第五卷
《报纸沙发》入选中国包装技术协会设计委员会《中国设计年鉴》第五卷
《果实般的鸟窝》入选中国包装技术协会设计委员会《中国设计年鉴》第五卷
《充气的灯》入选中国包装技术协会设计委员会《中国设计年鉴》第五卷
《奶牛壶》入选中国包装技术协会设计委员会《中国设计年鉴》第五卷
《弹簧体重计》入选中国包装技术协会设计委员会《中国设计年鉴》第五卷
《不锈钢弹片镜架眼镜》入选中国包装技术协会设计委员会《中国设计年鉴》第五卷
《弹簧镜腿眼镜》入选中国包装技术协会设计委员会《中国设计年鉴》第五卷

2004

《弹簧淋浴空间》获第十届全国美展银奖
《弹簧淋浴空间》入选意大利米兰2004年3月"SALONE DELL'ARREDO BAGNO 2nd"沙龙
《天蓝水清》入选第十届全国美展
《发光的书签》入选2004意大利“贝克啤酒”国际设计比赛
《社区猫粮站》获2004江苏省优秀工业设计奖
《椅子＋衣架》获2004江苏省优秀工业设计奖

2003

《弹簧镜腿眼镜》2003日本OPUS DESIGN AWARD国际眼镜设计比赛，入选
《不锈钢弹片镜架眼镜》2003日本OPUS DESIGN AWARD国际眼镜设计比赛，入选
《玻璃瓶眼镜》2003日本OPUS DESIGN AWARD国际眼镜设计比赛，入选
《宠物的日记》韩国“JAME2003国际互动”设计大赛，“挑战者”入选奖
《大龙纺织》中国包装技术协会设计委员会“中国设计之星”，商品类标识最佳设计奖
《墙与椅》入选深圳家具设计大赛

2002

《天蓝水清》法国“Young Designers,Looking into the Future”眼镜设计大赛，入选

《天蓝水清》日本名古屋“Design DO! 弱”国际设计大赛，入选

《滚动的软球》中央美术学院“为坐而设计”大赛，入选

2001

《风中的豆荚》2001 年无锡国际工业设计节青年设计师设计大赛，" 伊莱克斯杯 " 特别奖

《大嘴蛙沙发》2001 年无锡国际工业设计节青年设计师设计大赛，入选

《风中的豆荚》联合国教科文组织“持续与连接”国际设计大赛，获入选奖

《南京长江二桥桥铭牌》(合作)入选中国包装技术协会设计委员会《中国设计年鉴》第四卷

《折叠自行车》入选中国包装技术协会设计委员会《中国设计年鉴》第四卷

《利恒房产》标识入选中国包装技术协会设计委员会《中国设计年鉴》第四卷

2000

《禅宗指椅》2000 中国工业设计周“中国工业设计师设计大赛”，铜奖

《灯挂椅新译》入选深圳家具设计大赛

1999

《花瓶》联合国教科文组织“雅致中华”国际设计大赛，入选奖

《春兰摩托车》(集体)入选中国包装技术协会设计委员会《中国设计年鉴》第三卷

《新华电热水器》入选中国包装技术协会设计委员会《中国设计年鉴》第三卷

《富加乐》标识入选中国包装技术协会设计委员会《中国设计年鉴》第三卷

《同兴电器》标识入选中国包装技术协会设计委员会《中国设计年鉴》第三卷

《艺海影视》标识入选中国包装技术协会设计委员会《中国设计年鉴》第三卷

专利申请

2012 申请获批国家实用新型专利 48 项

2005-2006 申请获批国家外观专利 24 项

等待第三本